Henry Kiddle

A New Manual of the Elements of Astronomy

descriptive and mathematical

Henry Kiddle

A New Manual of the Elements of Astronomy
descriptive and mathematical

ISBN/EAN: 9783337275785

Printed in Europe, USA, Canada, Australia, Japan

Cover: Foto ©berggeist007 / pixelio.de

More available books at **www.hansebooks.com**

A

NEW MANUAL

OF THE

ELEMENTS OF ASTRONOMY,

DESCRIPTIVE AND MATHEMATICAL:

COMPRISING

THE LATEST DISCOVERIES AND THEORETIC VIEWS,

WITH DIRECTIONS FOR THE

USE OF THE GLOBES, AND FOR STUDYING THE CONSTELLATIONS.

BY

HENRY KIDDLE, A.M.,

ASSISTANT SUPERINTENDENT OF SCHOOLS, NEW YORK.

NEW YORK:
IVISON, BLAKEMAN, TAYLOR, & COMPANY,
138 & 140 GRAND STREET.
CHICAGO: 133 & 135 STATE STREET.
1871.

Electrotyped by SMITH & McDOUGAL, 82 and 84 Beekman St., N. Y.

PREFACE.

THIS work is designed to take the place of the "Manual of Astronomy," published by the author in 1852. In method of treatment it is entirely new, and is much more comprehensive than the previous work, containing a fuller exposition of elementary principles, and embodying the chief results of astronomical research during the last fifteen years, — a period exceedingly fruitful in discovery.

The *plan* is as far as possible *objective*, in a proper sense ; that is, it is based upon the conceptions of the pupil acquired by an actual observation of the phenomena of the heavens, to which his attention is constantly directed ; the relation between these phenomena and the facts inferred from them being clearly shown at every step. Great pains have been taken to divest that part of the subject which treats of the *Sphere* of its usual arbitrary and complex character by developing the requisite ideas before presenting formal definitions.

Simplified methods of computing the numerical elements, such as periods, distances, and magnitudes, are given throughout the work ; in most cases the calculations being made for the pupil, but without recourse to any other than elementary arithmetic and the most rudimental principles of geometry. These calculations are based on the recent determination of the solar parallax, and other elements as established by the latest observations and researches of distinguished astronomers ; and it is believed that, presented in this way, they will prove valuable as arithmetical exercises, as well as an important aid in imparting clear, correct, and permanent conceptions of the astronomical truths.

Brief historical sketches of the various discoveries—a most fascinating part of the subject—are given in connection with the facts to which they relate.   These, with all other matter designed to elucidate or exemplify the text, are printed in smaller type, and in distinct paragraphs, which have been distinguished by letters, so as to be readily referred to, and conveniently indicated by the teacher in assigning lessons.

The *Problems for the Globes* have been placed in connection with that part of the text to which they refer and in which they are designed to exercise the pupil.

The *illustrations* are copious, and have been engraved specially for this work, many from original drawings and diagrams ; the telescopic views, from drawings and photographs made by distinguished observers.   All the important English astronomical works of recent date have been carefully consulted.

The author hopes that this little volume, containing as it does a full exposition of the recent progress and present condition of the sublimest of all sciences, will prove a useful and acceptable addition to the educational facilities so copiously supplied at the present time.

New York, January 15, 1868.

CONTENTS.

INTRODUCTION.

CHAPTER I.

THE HEAVENLY BODIES—GENERAL PHENOMENA.

CHAPTER II.

THE PLANETS.

CHAPTER III.

MAGNITUDES OF THE SUN AND PLANETS.

CHAPTER IV.

ORBITAL REVOLUTIONS OF THE PLANETS.

CHAPTER V.

DISTANCES, PERIODIC TIMES, AND ROTATIONS OF THE PLANETS.

CHAPTER VI.

ASPECTS OF THE PLANETS.

CHAPTER VII.

THE EARTH.

CHAPTER VIII.

THE SUN.

CHAPTER IX.

THE MOON.

CHAPTER X.

ECLIPSES.

CHAPTER XI.

TIDES.

CHAPTER XII.

INFERIOR PLANETS.

CHAPTER XIII.

SUPERIOR PLANETS.

CHAPTER XIV.

MINOR PLANETS.

CHAPTER XV.

MUTUAL ATTRACTIONS OF THE PLANETS.

CHAPTER XVI.

COMETS.

CHAPTER XVII.

METEORS OR SHOOTING STARS.

CHAPTER XVIII.

THE STARS.

CHAPTER XIX.

NEBULÆ.

INTRODUCTION.

MATHEMATICAL DEFINITIONS.

1. ASTRONOMY * is that branch of science which treats of the heavenly bodies; as the sun, moon, stars, comets, etc.

Before even the simple elements of this science can be learned, it is necessary that the rudiments of geometry should be understood; hence, the following definitions are here presented as an introduction. For convenience of treatment, more are, however, inserted in this section than the student will need, at first, to apply. The teacher should, therefore, not only see that they are learned by way of preparation for the general subject, but be careful to recur to them, when the pupil reaches the parts of the subject to which they specially refer.

2. EXTENSION, or magnitude, may be measured in three directions; namely, length, breadth, and thickness. These are therefore called the *dimensions of extension.*

a. **Length** is the greatest dimension; **Thickness,** the shortest; **Breadth,** the other.

3. A LINE is that which is conceived to have only one dimension.

a. Lines have no real existence independently of extension, or solidity. They are purely abstract or imaginary quantities: the marks called lines are only *representatives* of them.

4. A STRAIGHT LINE is a line that does not change its direction at any point.

---

* Derived from the Greek words *Astron*, meaning *a star*, and *Nomos*, meaning *a law.*

---

5. A CURVE LINE is one that changes its direction at every point.

6. A POINT is that which is conceived to have no dimensions, but only position.

a. A point is represented by a dot (.).

b. A straight line measures the shortest distance between two points.

7. A SURFACE is that which is conceived to have two dimensions, length and breadth.

8. A PLANE SURFACE, OR PLANE, is a surface with which, if a straight line coincide in two points, it will coincide in all.

a. That is, a straight line cannot lie partly in a plane, and partly out of it ; and if applied to it in any direction, it will coincide with it throughout its whole extent. The term *plane* does not imply any limitation, or boundary, but signifies indefinite direction, without change, both as to length and breadth.

9. A plane bounded by lines is called a PLANE FIGURE.

Fig. 1.

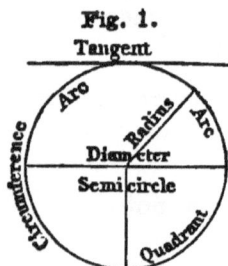

10. A CIRCLE is a plane figure bounded by a curve line every point of which is equally distant from a point within, called the *centre*.

11. The curve line that bounds a circle is called the CIRCUMFERENCE.

12. The DIAMETER of a circle is a straight line drawn through its centre from one point of the circumference to another.

13. The RADIUS is a straight line drawn from the centre to the circumference.

14. An ARC is any part of the circumference.

15. A TANGENT is a line which touches the circumference in one point.

16. A SEMICIRCLE is one-half of a circle; a QUADRANT is a quarter of a circle.

17. The circumference of a circle is supposed to be divided into 360 degrees, each degree into 60 minutes, and each minute into 60 seconds.

a. Degrees are marked (°); minutes, ('); and seconds, ('').

18. An ANGLE is the difference in direction of two straight lines that meet at a point, called the vertex.

Fig. 2.

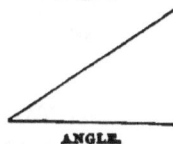

ANGLE.

a. It is of the greatest importance that the student of Astronomy should form a clear idea of an angle, since nearly the whole of astronomical investigation is based upon it. The apparent distance of two objects from each other, as seen from a remote point of view, depends upon the *difference of direction* in which they are respectively viewed; that is to say, the angle formed by the two lines conceived to be drawn from the objects, and meeting at the eye of the observer. This is called the *angular distance* of the objects, and, as will readily be understood, increases as the two objects depart from each other and from the general line of view.

Fig. 3.

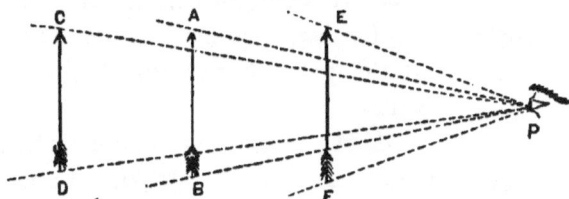

19. The ANGLE OF VISION, or VISUAL ANGLE, is the

angle formed by lines drawn from two opposite points of a distant object, and meeting at the eye of the observer.

a. It will be easily seen that, as the *apparent size* of a distant object depends upon the angle of vision under which it is viewed, it must diminish as the distance increases, and *vice versa.*

Thus, the object A B (Fig. 3) is viewed under the angle A P B, which determines its apparent size in that position; but when removed farther from the eye, as at C D, the angle of vision becomes C P D, an angle obviously smaller than A P B, and hence the object appears smaller. At E F, the object appears larger, because the visual angle E P F is larger. The farther the object is removed, the less the divergence of the lines which form the sides of the angle; and the nearer the object is brought to the eye, the greater the divergence of the lines.

20. An angle is measured by drawing a circle, with the vertex as a centre, and with any radius, and finding the number of degrees or parts of a degree, included between the sides.

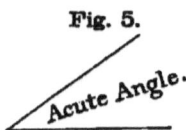

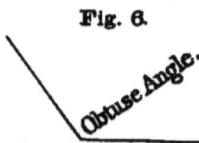

Fig. 4.

Right Angle.

Fig. 5.

Acute Angle.

Fig. 6.

Obtuse Angle.

21. A RIGHT ANGLE is one that contains 90 degrees, or one-quarter of the circumference.

22. When one straight line meets another so as to form a right angle with it, it is said to be PERPENDICULAR.

23. A straight line is said to be perpendicular to a circle when it passes, or would pass if prolonged, through the centre.

24. An angle less than a right angle is called an ACUTE ANGLE; one greater than a right angle is called an OBTUSE ANGLE.

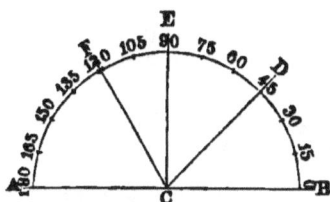
Fig. 7.

In the annexed diagram, the semi-circumference is used to measure all the angles having their vertices, or angular points, at C. Thus B C D, containing the arc B D, is an angle of 45°; B C E, an angle of 90°; and B C F, of 120°. The points A and B are at the angular distance of 180°, or two right angles from each other.

25. A TRIANGLE is a plane figure bounded by three sides.

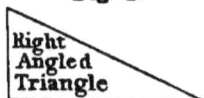
Fig. 8.

Right Angled Triangle

a. The sum of the three angles of every triangle is equal to two right angles.

b. A triangle that contains a right angle is called a *Right-angled Triangle.*

Fig. 9.

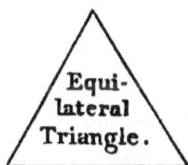

Equilateral Triangle.

c. A triangle having equal sides is called an *Equilateral Triangle.*

d. Each of the angles of an equilateral triangle is an angle of 60°; since the three angles are equal to each other, and their sum is equal to 180°.

Fig. 10.

Parallel Lines.

26. PARALLEL LINES are those situated in the same plane, and at the same distance from each other, at all points.

Fig. 11.

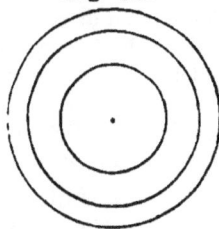

a. Parallel lines may be either straight or curved.

b. The circumferences of concentric circles, that is, circles drawn around the same centre, are parallel.

27. AN ELLIPSE is a curve line, from any point of which if straight lines be drawn to two points within, called the foci, the sum of these lines will be always the same.

QUESTIONS.—25. What is a triangle? *a.* Sum of its three angles? *b.* What is a right-angled triangle? *c.* An equilateral triangle? *d.* The value of each of its angles? Why? 26. What are parallel lines? *a.* Are they always straight? *b.* When are circles parallel? 27. What is an ellipse?

Fig. 12.

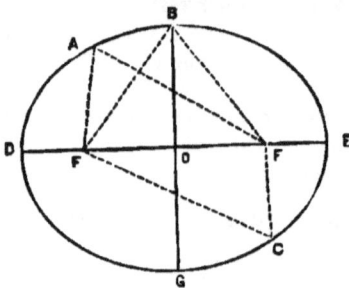

The curve line D B E G represents an ellipse, the sum of the two straight lines drawn to F and F, the foci, from the points A, B, and C, respectively, being always equal. This sum is equal to the longest diameter, D E.

28. The longest diameter of an ellipse is called the MAJOR* AXIS; † and the shortest diameter, the MINOR* AXIS.

In the diagram, D E is the major axis, and B G the minor axis.

29. The distance from either of the foci to the centre of the ellipse is called the ECCENTRICITY ‡ of the ellipse.

a. It will be readily seen that the greater the eccentricity of an ellipse, the more elongated it is, and the more it differs from a circle; while, if the eccentricity is nothing, the two foci come together, and the ellipse becomes a circle.

b. The distance from the extremity of the minor axis to either of the foci is always equal to one-half of the major axis.

In the above diagram, F O is the eccentricity, and B F is equal to D O. The amount of eccentricity of any ellipse is ascertained by comparing it with one-half the major axis. Thus, in the diagram, O F being about one half of O D, the eccentricity of the ellipse may be nearly expressed by .5.

30. A SPHERE, or GLOBE, is a round body every point of the surface of which, is equally distant from a point within, called the centre.

31. A HEMISPHERE is a half of a globe.

32. The DIAMETER of a sphere is a straight line drawn

---

* *Major* and *Minor* are Latin words, meaning *greater* and *less*.
† A Latin word meaning an *axle*, that on which any thing turns.
‡ From the Greek, *ec*, from, and *centron*, the centre.

QUESTIONS.—28. What is the major axis? The minor axis? 29. What is meant by eccentricity? *a.* What does it show? *b.* Distance of foci from the extremity of minor axis? Explain from the diagram. 30. What is a sphere? 31. A hemisphere? 32. The diameter of a sphere?

through the centre, and terminated both ways by the surface of the sphere.

33. The RADIUS of a sphere is a straight line drawn from the centre to any point of the surface.

34. Circles drawn on the surface of a sphere are either GREAT CIRCLES or SMALL CIRCLES.

35. GREAT CIRCLES are those whose planes divide the sphere into equal parts.

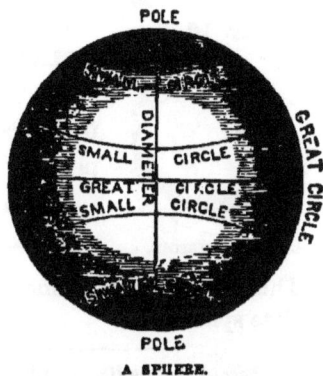

Fig. 13.

A SPHERE.

36. SMALL CIRCLES are those whose planes divide the sphere into unequal parts.

37. The POLES OF A CIRCLE are two opposite points on the surface of the sphere, equally distant from the circumference of the circle.

a. The poles of a great circle are, of course, 90° distant from every point of its circumference.

b. Two circles of the sphere are parallel when they are equally distant from each other at every point.

c. Two circles are perpendicular to each other when their planes are perpendicular, or at right angles with each other.

d. The Plane of a Circle or any other figure is the indefinite plane surface on which it may be conceived to be drawn.

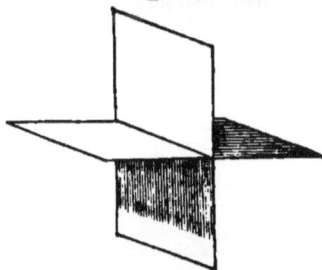

Fig. 14.

PERPENDICULAR PLANES.

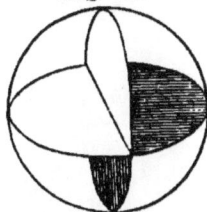

Fig. 15.

PLANES OF GREAT CIRCLES.

QUESTIONS.—33. What is the radius? 34. How are circles of the sphere divided? 85. What are great circles? 36. Small circles? 37. Poles of a circle? *a.* Poles of a great circle? *b.* When are circles of the sphere parallel? *c.* When perpendicular? *d.* What is meant by the plane of a circle?

38. A SPHEROID* is a body resembling a sphere.

Fig. 16.

39. There are two kinds of spheroids; OBLATE and PROLATE Spheroids.

40. An OBLATE SPHEROID is a sphere flattened at two opposite points, called the poles.

41. A PROLATE SPHEROID is a sphere extended at two opposite points.

OBLATE SPHEROID.

Thus, an orange is a kind of oblate spheroid; and an egg, a kind of prolate spheroid.

Spheroid means *like* a sphere. *Oid* is from the Greek word *eido*, meaning *to resemble*.

QUESTIONS.—38. What is a spheroid? 39. How many kinds of spheroids? 40. What is an oblate spheroid? 41. A prolate spheroid?

CHAPTER I.

1. The SUN, the MOON, and the STARS, are the most conspicuous bodies of which astronomy treats.

a. **Antiquity of the Science.**—The various appearances presented by these bodies must always have engaged the attention of mankind. The sublime spectacle of the starry heavens would naturally, in the earliest times, excite the admiration of the most careless or ignorant observer; and the curiosity of mankind would be early aroused to ascertain the nature of those "refulgent lamps" which lend so much splendor and beauty to the otherwise sombre gloom of night.

Hence, we find that astronomy is a very ancient science. The shepherds of Chaldea,* and the priests of Egypt and India, had, in remote antiquity, made some progress in astronomical discovery; and, it is said, Chinese observations are on record that date back more than 1000 years before Christ.

b. **Ordinary Phenomena.**—The most obvious phenomena † connected with these bodies are their rising and setting, and their constant motion in the same general direction from one side of the heavens to the other; and consequently these appearances were probably among the first that incited to scientific inquiry.

c. **The Earth's Rotation.**—The simple fact that the earth—the body on which we are placed—turns round once every twenty-four hours, clears away all difficulty in explaining these daily appearances; but it was not until comparatively recent times that mankind could be brought generally to accept this truth.

As late as 1633, it was deemed irreligious to believe in the motions

---

* A country of antiquity, situated between the Euphrates and Tigris rivers.

† *Phenomena*, plural of *phenomenon*, a Greek word meaning *an appearance.*

QUESTIONS.—2. Which are the most conspicuous of the heavenly bodies? *a.* Astronomy—why an ancient science? *b.* What are the most obvious phenomena? *c.* The earth's rotation—what does it explain? Galileo?

of the earth ; and Galileo, in his seventieth year, was imprisoned, and finally compelled to acknowledge himself as guilty of error and heresy in teaching this astronomical truth.

d. **The Stars** appear to keep very nearly the same situations with respect to each other ; and hence were called *Fixed Stars*, to distinguish them from other bodies which resemble, in their general appearance, stars, but seem to move about in the heavens, at one time being near one star, then another, now moving in one direction, then in another; thus, as it were, wandering about in the heavens. For this reason, such bodies were called *planets*, or *wandering stars*.

[In the Greek language, *planētes* means a wanderer. The term *fixed stars* is now but little used by astronomers ; and in this work, when the term *stars* is used, it is intended to designate fixed stars.]

2. The apparent motions of the sun, planets, and stars are explained by supposing, 1. That the earth is a sphere, or nearly so ; 2. That it turns on its axis ; 3. That the earth and planets revolve around the sun ; and, 4. That the stars are situated at an immense distance from the sun and planets, in the regions of space,—a distance so vast that their movements with respect to each other can not generally be discerned.

3. The sun with all the bodies revolving around it, is called the SOLAR * SYSTEM.

a. **Copernican System.**—This arrangement of the sun in the centre with the planets revolving around it, is sometimes called the *Copernican System*, from Nicholas Copernicus, who, in 1543, revived the doctrine taught by Pythagoras, a Greek philosopher, more than 2,000 years before, that the sun is the central body, and that the earth and planets revolve around it.

b. **Ptolemaic System.**—Previous to Copernicus, the general belief, for more than two thousand years, had been that the earth is the cen-

---

* From the Latin word *Sol*, meaning *the sun*.

QUESTIONS.—*d.* Stars and planets—how distinguished? What does *planet* mean? 2. How are the apparent motions of the sun, planets, and stars explained? 3. What is the solar system? *a.* The Copernican system—why so called? Pythagoras? *b.* The Ptolemaic system—why so called? Describe it.

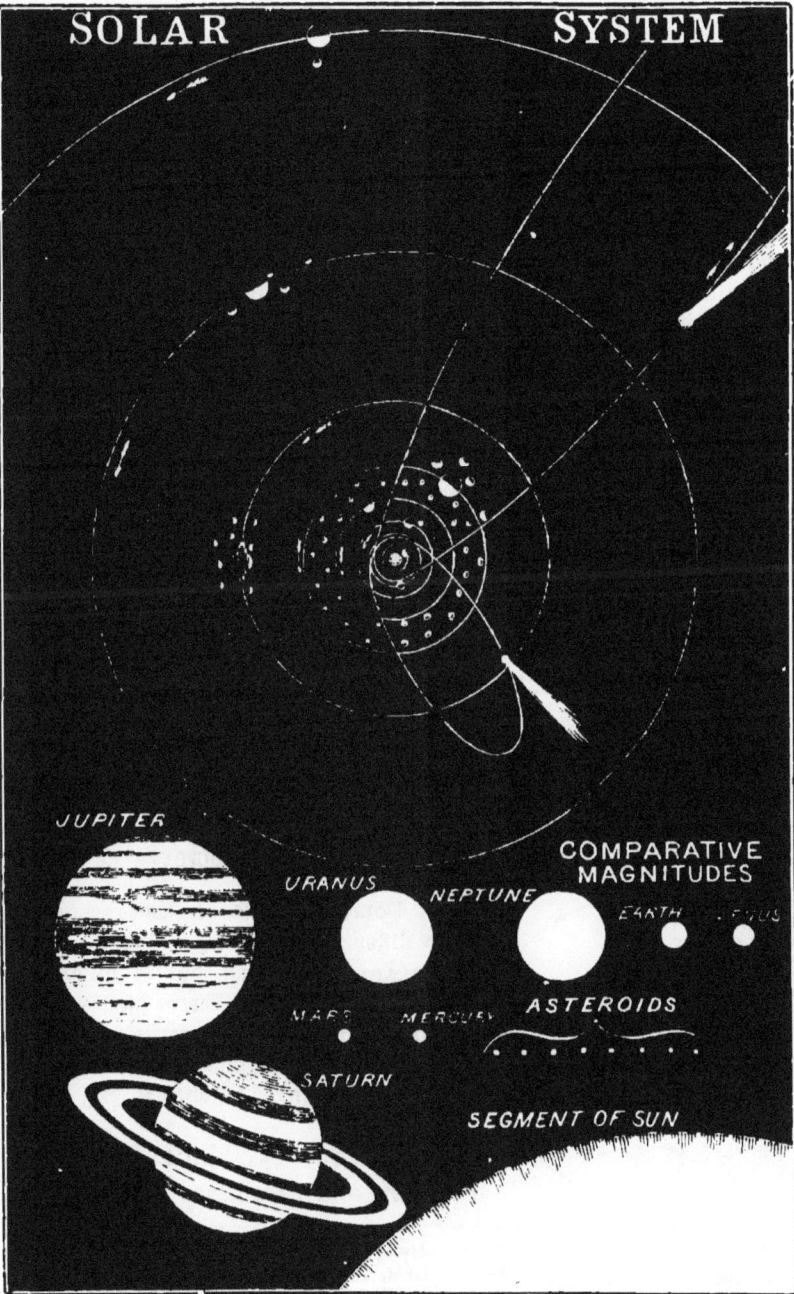

SOLAR SYSTEM

JUPITER

URANUS NEPTUNE

COMPARATIVE
MAGNITUDES

EARTH VENUS

MARS MERCURY ASTEROIDS

SATURN

SEGMENT OF SUN

tre of the universe, and that all the other bodies revolve around it, in the following order : the moon, then the sun and planets in their order, and then the stars. Each of these bodies was conceived by Aristotle to be set in a hollow, crystalline sphere, perfectly transparent, by which it was carried around the earth and prevented from falling upon it. This celebrated system is very ancient, but being advocated and illustrated by Ptolemy, an eminent astronomer who flourished at Alexandria, in Egypt, about 140 A.D., it was subsequently called the *Ptolemaic System.*

c. **The Invention of the Telescope.**—The doctrine of Copernicus, as promulgated in his great work, styled the " Revolutions of the Celestial Orbs," published in 1543, was at first generally rejected, and despised as visionary and absurd ; but the invention of the telescope, in 1610, and the discoveries made by means of it, by Galileo and others, afforded abundant evidence of the truth of this hypothesis.

Fig. 17.

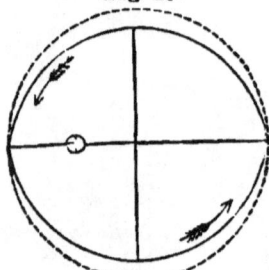

A PLANET'S ORBIT.

Fig. 18.

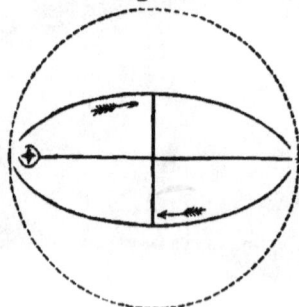

A COMET'S ORBIT.

4. PLANETS are bodies that revolve around the sun.

5. The path in which we may conceive a planet to revolve is called its *orbit.*

6. The planets all revolve around the sun in the same direction, in orbits nearly circular, and situated in nearly the same plane.

a. **Comets' Orbits.**—In these respects they differ from *Comets,* which also revolve around the sun, but in different directions, in widely different planes, and in very elongated, or eccentric orbits ; that is, elliptical orbits of great eccentricity. [See Introduction, Art. 29.]

7. There are two kinds of planets ; *Primary* and *Secondary Planets.*

8. PRIMARY PLANETS are those that revolve around the sun only.

9. SECONDARY PLANETS, generally called SATELLITES,* are those that revolve around their primaries, and, with them, around the sun.

The moon is an example of a secondary planet. It is the earth's satellite, its revolution around the earth being clearly indicated by the changes which it undergoes each month.

10. The Solar System is thus composed of the sun, the primary planets, the secondary planets, and the comets; while the stars are bodies situated at an immense distance beyond the system.

11. All the heavenly bodies may be divided into two general classes; namely, LUMINOUS BODIES and OPAQUE BODIES.

12. Luminous bodies are such as shine by their own light; opaque bodies are such as shine by reflecting the light of some luminous body.

a. The sun is evidently a luminous body ; for we receive from it both light and heat, and in every position it presents a resplendent circular surface, called its *Disc.* The moon is as evidently opaque, since it does not always exhibit an entire disc, but various portions of it at different times, such portions being called *Phases.*

b. The stars present no disc, but only luminous points, shining with that *twinkling light* which indicates their intense brilliancy and vast distance. They are believed to be luminous bodies, since we can discover no body from which they could receive their light ; and, moreover, the light itself which they emit has different properties from those possessed by reflected light.

c. The planets, though in general appearance resembling the stars, may readily be distinguished from them by their *steady light.* When viewed through a telescope, some of them exhibit phases like those of the moon.

---

* From the Latin word *satelles,* (plural, *satellites,*) meaning a guard.

QUESTIONS.—8. What are primary planets? 9. Secondary planets? 10. Of what is the solar system composed? Where are the stars situated? 11. What general division of the heavenly bodies? 12. What are luminous bodies? Opaque bodies? *a.* Proof that the sun is luminous? That the moon is opaque? *b.* Proof that the stars are luminous? *c.* Light of planets—how distinguished from that of stars?

CHAPTER II.

THE PLANETS.

13. There are eight large primary planets in the solar system, besides a great number of smaller ones, called MINOR PLANETS, or ASTEROIDS.*

14. The names of the eight large primary planets, in the order of their distances from the sun, are Mercury, Venus, the Earth, Mars, Jupiter, Saturn, Uranus, and Neptune.

15. All the primary planets except the earth are divided into two classes, INFERIOR and SUPERIOR PLANETS.

16. Mercury and Venus are called inferior planets, because they revolve within the orbit of the earth; Mars, Jupiter, Saturn, Uranus, and Neptune are called superior planets, because they revolve beyond the orbit of the earth. Instead of the terms inferior and superior, *interior* and *exterior* are sometimes used.

a. **Vulcan.**—A planet inferior to Mercury has been supposed to exist; and in 1859, a French astronomer was thought by some to have discovered it. Later observations have not, however, confirmed, but rather disproved, its existence. The name given to this supposed planet is *Vulcan.*

17. The MINOR PLANETS are very small planets which revolve around the sun, between the orbits of Mars and Jupiter. *Ninety-six* have been discovered (1868).

a. These small planets were at first, and have been, very generally, called **Asteroids**; they have also been called **Planetoids**. The name

---

* From the Greek *aster*, meaning *a star*, and *eido*, to resemble.

QUESTIONS.—13. How many primary planets in the solar system? 14. What are the names of the large planets? 15. How divided? 16. Which are called inferior? Which superior? Why? *a.* Vulcan—what is said of it? 17. What are the minor planets? Their number? *a.* What other names are applied to them?

above given has, however, been extensively used by astronomers, and appears to be the most significant and appropriate.

18. All the primary planets revolve around the sun in the direction which is designated *from west to east.*

a. It is difficult to fix definitely the direction of circular motion, since when viewed in one position it may seem to be from left to right (as the hands of a clock move), and in another, from right to left. The motion of the planets, as we view them, is from *right to left,*—the reverse direction of the hands of a clock. This is the direction indicated in the diagrams of this work.

b. East is at or near where the sun rises; West, at or near where it sets. South is in the direction of the sun's place at noon; North, directly opposite the south. If we stand so as to face the north, the south will be behind us, the east on the right hand, and the west on the left.

19. Secondary planets, or satellites, have two motions: one around their primaries, and another, with them, around the sun.

20. Eighteen satellites are known to exist in the Solar System: the earth has one, called the moon; Jupiter has four; Saturn, eight; Uranus, four; and Neptune, one.

a. While Uranus is undoubtedly attended by at least four satellites, there is a very great uncertainty as to the exact number which belong to it. Sir William Herschel, by whom this planet was discovered, in the latter part of the last century, detected, as he thought, six satellites; but only *two* of these have been observed by other astronomers, which, with two others discovered by Lassell in 1853, make the four referred to in the text.

21. The satellites of the earth, Jupiter, and Saturn revolve around their primaries from west to east; those of Uranus and Neptune, from east to west.

a. With the exception of the satellites of Uranus, and the satellite of Neptune, all the planets of the Solar System revolve in the same direction, that is, from *west to east.*

QUESTIONS.—18. What is the direction of the planets' motion? *a.* How defined? *b.* What is meant by *East? West? South? North?* 19. What motions have satellites? 20. How many are known to exist? Enumerate them? *a.* Satellites of Uranus? 21. In what direction do the satellites revolve? *a.* What uniformity of motion in the solar system?

CHAPTER III.

MAGNITUDES OF THE SUN AND PLANETS.

22. The SUN is by far the largest body in the Solar System, being more than 500 times as large as all the planets taken together.

23. Its diameter is a little more than 850,000 miles.

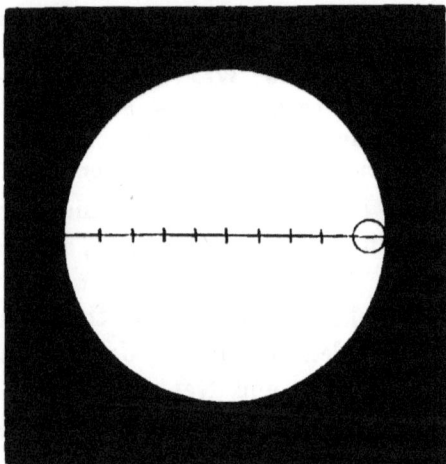

Fig. 19

COMPARATIVE SIZE OF SUN AND JUPITER.

24. The LARGEST PLANET is Jupiter, its diameter being 85,000 miles, or one-tenth as large as that of the sun.

a. Volume, Mass, Density.—The diameter of Jupiter being one-tenth as large as the sun's, its *volume*, or *bulk*, is one one-thousandth (.001) that of the sun ; since it is only one-tenth as great in each dimension,—length, breadth, and thickness ; and $\frac{1}{10} \times \frac{1}{10} \times \frac{1}{10} = \frac{1}{1000}$.

This is expressed generally by saying that *solid bodies of similar shape are in proportion to the cubes of their like dimensions.*

b. The *volume* of a body is the amount of space which it occupies,

---

QUESTIONS —22. What is the comparative size of the sun ? 23. Length of its diameter ? 24. Which is the largest planet ? Its diameter ? *a.* Comparative volume of the sun and Jupiter ? How found ? *b.* Define *volume.*

as indicated by its length, breadth, and thickness. The volumes of bodies are in proportion to the products of their three dimensions.

c. Two bodies may be equal in volume, but contain very different quantities of matter, owing to the different degrees of compactness of their substance. Thus, a piece of cork, equal in bulk to a piece of lead, contains only about $\frac{1}{11}$ as much matter. The quantity of matter which a body contains is called its *mass;* the degree of compactness of its subtance is called its *density.*

d. The mass of a body depends upon its volume and density considered conjointly. Thus, if the volumes of two bodies are as 2 to 3, and their densities, as 1 to 5, their masses will be as 1 × 2 to 3 × 5, or as 2 to 15; that is to say, the mass of the second will be 7½ times as great as that of the first.

25. The following are the diameters of the large primary planets in miles:—

[These are given in round numbers so as to be easily remembered; a more exact statement will be found in another part of this work. It is important that the student should carefully commit to memory these numbers, since the relative magnitudes of the planets form the basis of much of the reasoning in respect to the solar system.]

1. Jupiter, . . 85,000.	5. Earth, . . 7,912.
2. Saturn, . . 70,000.	6. Venus, . . 7,500.
3. Neptune, . 37,000.	7. Mars, . . 4,300.
4. Uranus, . . 33,000.	8. Mercury, . 3,000.

a. **Major and Terrestrial Planets.**—The first four of these planets, it will be seen, are very much larger than the remaining four, and are, for this reason, sometimes called the *Major Planets;* while the others, being in the vicinity of the earth, are sometimes called the *Terrestrial Planets.*

b. **Illustration.**—A clear idea of the comparative size of the sun and planets may be obtained by conceiving the sun to be a globe two feet in diameter. Mercury and Mars would then be of the size of pepper-corns; the earth and Venus, of the size of peas; Jupiter and

Saturn, as large as oranges; and Neptune and Uranus, as large as full-sized plums. [See figure, page 19.]

26. The MINOR PLANETS, or ASTEROIDS, are all of very small size, the diameter of the largest not exceeding 300 miles, and that of the smallest being only a very few miles.

a. **Entire Mass of the Minor Planets.**—It has been computed by the celebrated French mathematician, Le Verrier, that their entire mass, however many may exist, can not exceed one-fourth that of the earth. This calculation is based on the amount of disturbance occasioned by their united attraction, in the motions of the Earth and Mars. Now, the diameter of the largest being only about $\frac{3}{80}$ of the diameter of the earth, its volume must be only $\frac{1}{19000}$ of the earth's; and hence, it would require 4,750 planets as large as the largest of the asteroids to equal the amount specified by the mathematician.

27. All the SATELLITES are smaller than Mars, and with the exception of two, one of Jupiter's and one of Saturn's, are smaller than Mercury.

a. The diameter of the moon is 2,160 miles; the four satellites of Jupiter, excepting one, are larger than the moon; and the eight satellites of Saturn, excepting one, are smaller than the moon.

28. The following presents a comparative view of the densities of the primary planets, as compared with that of water:

1. Mercury, . . $6\frac{1}{3}$.		5. Jupiter, . . $1\frac{3}{8}$.	
2. Venus, . . $5\frac{7}{8}$.		6. Uranus, . . 1.	
3. Earth, . . . $5\frac{5}{8}$.		7. Neptune, . $\frac{9}{10}$.	
4. Mars, ... 4.		8. Saturn, . . $\frac{3}{4}$.	

a. It will be seen that the terrestrial planets are all of considerably greater density than the major planets; and that the densities diminish, with the exception of Saturn, as the distance of the sun increases.

29. The density of the sun is only one-fourth that of the earth, or about $1\frac{1}{2}$ that of water. The density of the moon is about $3\frac{1}{2}$ that of water.

a. **Masses of the Planets.**—If we arrange the planets according to their mass, they will stand in precisely the same order as when arranged according to volume, though not in the same proportion. The student can verify this by applying the principle explained in Art. 3, *d.* The following table presents a general view of the comparative masses of the sun and planets, expressed in approximate numbers, the earth being 1 :—

SUN, . . . 315,000.

MAJOR PLANETS.	Jupiter, . . 301.		TERRESTRIAL PLANETS.	Earth, . . 1.	
	Saturn, . . 90.			Venus, . . $\frac{1}{4}$.	
	Neptune, . 16$\frac{1}{2}$.			Mars, . . $\frac{1}{18}$.	
	Uranus, . . 12$\frac{1}{2}$.			Mercury, . $\frac{1}{16}$.	

Moon, . . . $\frac{1}{80}$.

QUESTIONS.—29. What is the density of the sun and moon? *a.* Comparative masses of the planets?

CHAPTER IV.

THE ORBITAL REVOLUTIONS OF THE PLANETS.

30. A planet's revolution in its orbit is sustained by the united action of two forces; namely, the *Centripetal** and the *Centrifugal* † forces.

a. **Laws of Motion.**—No portion of matter can set itself in motion; nor, when in motion, can it stop itself. Whatever sets a body in motion, or stops it when in motion, is called **Force.**

b. A body when acted upon by a single force, moves in a straight line; and will continue to move in the same direction, and with the same velocity, until acted upon by some other force.

Fig. 20.

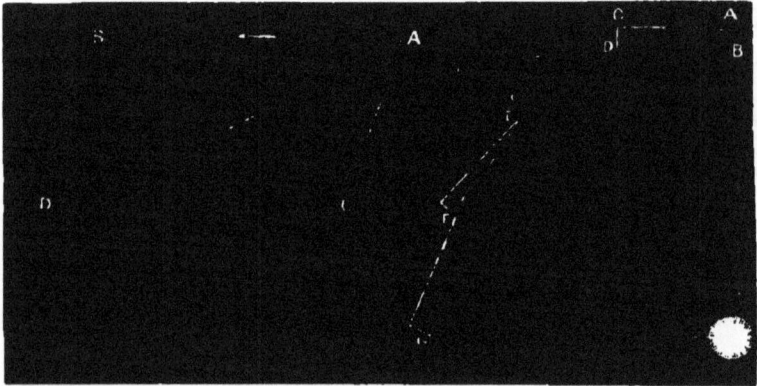

* From the Latin words *centrum*, meaning *the centre*, and *peto*, meaning *to seek.*

† From the Latin words *centrum*, and *fugio*, meaning *to flee from.*

c. A force may be either *impulsive,* that is, acting once and then ceasing to act, or *continuous,* that is, acting constantly.

d. **Resultant Motion.**—When a body is impelled by two forces in different, but not opposite, directions, it moves in a straight line between them. This line is the diagonal of a parallelogram of which the lines that represent the two forces are adjacent sides.

Thus, let A B (Fig. 20.) represent the line over which the body A would pass in a certain time under the influence of one force, and A C, the line over which it would pass in the same time, if acted upon by another force; then under the simultaneous action of both forces, it will pass over the line A D in the same time, and continue to move in this line until acted upon by some third force. This line is called the *resultant* of the two forces.

e. **Curvilinear Motion.**—If one of the two forces were a continuous force, the body would be drawn, at every point, from the straight line, and, consequently, would move in a curve line; and these two forces might be so related to each other that the body would move around the centre of the continuous force in a circle or ellipse. In that case, the continuous force would be the *Centripetal Force,* and the impulsive force, the *Centrifugal.*

31. The CENTRIPETAL FORCE is that by which a body tends to approach the centre, or point around which it is revolving.

32. The CENTRIFUGAL FORCE is that by which a body tends to fly off from the orbit in which it is revolving.

33. The centripetal force which acts upon the primary planets is the attraction of the sun; that which acts upon the secondary planets is the attraction exerted by their respective primaries.

34. *All bodies attract each other in direct proportion to the mass, or quantity of matter, and inversely as the square of the distance.* That is, a body containing twice the quantity of matter of another body exerts twice the force; but, at twice the distance, would exert only one-fourth the force.

a. **Newton's Discovery.**—This is the celebrated *law of universal gravitation* discovered by Sir Isaac Newton, in 1665. It is said to have been suggested to his mind by the simple occurrence of an apple's falling from a tree. Observing that all bodies, when unsupported, fall toward the centre of the earth, he inferred that this must be occasioned by an attractive force exerted by the earth ; and from this, his mind was led to inquire whether it is not the same force, that is, a force acting according to the same law, which confines the moon in her orbit around the earth, and the earth and planets in their orbits around the sun. The calculations which he made proved that these conjectures were correct, and thus established the law.

b. **Centrifugal Force.**—The centrifugal force must arise from an impulse originally given to the planets when they commenced their motions ; since, without such an impulse, they would have simply moved toward the sun and have been incorporated with it. And if the centrifugal force were now destroyed, the planets would all move in straight lines to the sun ; while, if the attraction of the sun were suspended, they would move off into space in *tangent lines* to their orbits.

Let A B (Fig. 20) represent the amount of the centripetal, and A C that of the centrifugal force, for a given time; then completing the parallelogram, and drawing the diagonal A D, we find the point which the body when acted on by both forces will reach in that time. E, F, and G may be shown in a similar way to be the points reached by the body at the end of successive periods of time of an equal length ; and thus, if the forces acted by impulses, the body would describe the broken line formed by the diagonals of the parallelograms ; but as the force of gravitation is a continuous force, the revolving body describes a curve, which may either be a circle or an ellipse.

35. The planets' orbits are ellipses, having the sun or central body in one of the foci.

a. **Kepler's Laws.**—This is the first of the three celebrated truths pertaining to the planetary motions, discovered by Kepler after many years of investigation, and announced by him in 1609 ; hence, called " Kepler's Laws." Previous to this time, the general belief among astronomers had been that the planets' orbits are circular in form, since they conceived the circle to be the most perfect and beautiful of curves; but, according to this theory, they had found very great difficulty in accounting for the irregularities in the apparent motions of the planets.

b. **The Epicycle.***—This was, however, partially accomplished by ingeniously supposing that the planet, instead of revolving in a simple orbit, revolved in a small circle, called an epicycle, the centre of which moved around in a circular orbit. This hypothesis was invented, it is supposed, about two centuries B. C., and was adopted by Ptolemy and all the great astronomers, including Copernicus himself, who could not account for the apparent irregularities in the motions of Mars, which has a very eccentric orbit, on any other hypothesis. The explanation by the epicycle is illustrated by the annexed diagram.

The small circle represents the epicycle, the centre of which moves in the large circle around the sun, S. At A the planet is nearest to the sun; but while it performs one-quarter of a revolution in the epicycle, the latter also moves over one-quarter of its orbit, and thus the planet is carried to B, and in a similar manner to C, its farthest point from the sun, and thence through D to A again. The difference between its greatest distance from the sun C S, and its least distance A S, is equal to the diameter of the epicycle.

Fig. 21.

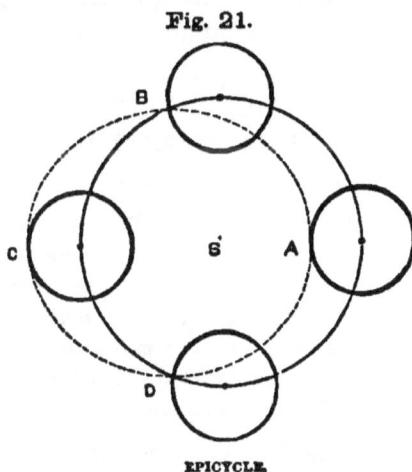

EPICYCLE.

c. **Tycho Brahe.**—Such ingenious but cumbrous hypotheses could only be sustained by the most imperfect observations made with the rudest instruments; but when astronomy, as an art of observation, came to be cultivated, they were necessarily exploded. Tycho Brahe is justly to be considered the founder of modern practical astronomy. He was born in 1546, in Sweden, and so great a reputation did he

---

* From the Greek words *epi*, meaning *upon*, and *cycle*, a circle; that is, *a circle upon a circle.*

QUESTIONS.—*b.* What is the hypothesis of the epicycle? Explain by the diagram. *c.* Tycho Brahe? Value of his labors, and use made of them by Kepler?

acquire, that Ferdinand, king of Denmark, built for him, on an island
at the mouth of the Baltic, a magnificent observatory, which he styled
"Uraniberg. or the City of the Heavens." His accurate observations
of the planets were the means of conducting Kepler to the discovery
of his famous laws. No less than nineteen different hypotheses were
made by Kepler, before he could bring his mind to abandon the theory
of the circular motion of the planets, and then he assumed the ellipse,
as being the next most beautiful curve. The adoption of this hypothesis at once reconciled the *computed* with the *observed* place of Mars ;
and, on applying it to the other planets, he found still more convincing
proof of its truth.

36. The straight line that joins the sun or central body
with the planet at any point of its orbit, is called the RADIUS-
VECTOR.*

37. The point of a planet's orbit nearest to the sun is
called its PERIHELION ; † the point farthest from the sun, its
APHELION.‡

a. Apsides.—The aphelion and perihelion are, of course, the extremities of the major axis. These two points are sometimes called the
Apsides,§ and the line that joins them, the *Line of Apsides.*

b. One-half of the sum of the aphelion and perihelion distances of a
planet is, of course, the *mean distance.* This is always equal to the
distance of a planet from the sun when it is at either extremity of its
minor axis. [See Introduction, Art. 29, *b.*]

38. The radius-vector of a planet's orbit passes over equal
spaces in equal times. This is the second of Kepler's laws.

If S (Fig. 22) represent the sun in the focus of a planet's elliptical orbit, A
will be the aphelion, P the perihelion, and A S, B S, C S, etc., the radius-

---

* *Vector,* in the Latin, means *that which carries.* The radius-vector is conceived to carry the planet as it moves around in its orbit.

† From the Greek *peri*, meaning around or *near ;* and *helios,* the sun.

‡ From *apo,* meaning *from,* and *helios. Apo* in combination becomes *aph.*

§ *Apsis,* plural *apsides,* is from the Greek, and means a *joining.*

QUESTIONS.—36. Define *radius-vector.* 37. Define *perihelion* and *aphelion.* *a.* Apsides and apsis line. *b.* What is mean distance? 39. What is Kepler's second law?
Explain from the diagram.

vector in different positions of the planet. The planet moves in its orbit so
that the spaces A S B, B S C, etc., may
be equal, if described in equal times.
It has therefore to move much faster
in the perihelion than in the aphelion,
since at the former point the spaces
must be wider in order to make up
for their diminished length.

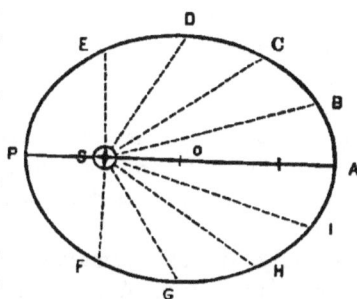

Fig. 22.

ELLIPTICAL ORBIT.

a. Orbital Veloctiy.—The velocity of a planet must therefore be
variable when it moves in an elliptical orbit, being greatest at the
perihelion, least at the aphelion, and alternately increasing and diminishing between these points.

b. The second law of Kepler is equally true for every kind of orbit,
including circular orbits ; but in the latter, the radius of the circle
would be the radius-vector, and not only would the spaces described
be equal, but also the different portions of the orbit, and consequently,
the velocity would be uniform. The orbits of the satellites of Jupiter
and Uranus are almost, if not exactly, circular.

**39. The squares of the periodic times of the planets are in
proportion to the cubes of their mean distances from the
sun, or central body.**

a. That is to say, if we square the times which any two planets
require to complete a revolution around the sun, and then cube their
mean distances, the ratio of the squares will be equal to that of the
cubes. This law applies to the *secondary* as well as the primary
planets.

b. History.—This is the third and most celebrated of Kepler's laws.
It establishes a most beautiful harmony in the Solar System. In his
work on "Harmonics," Kepler first made it known, with a perfect
burst of philosophic rapture. "What I prophesied, twenty-two years
ago,—that for which I have devoted the best part of my life to astronomical contemplations,—at length I have brought to light, and have

QUESTIONS.—*a.* Velocity of a planet—when variable? *b.* When uniform? 39. Relation of periodic times to distances? *a.* Is it true of the satellites? *b.* History of its
discovery? (Repeat the three laws of Kepler.)

recognized its truth beyond my most sanguine expectations. It is now eighteen months since I got the first glimpse of light, three months since the dawn ; very few days since the unveiled sun, most admirable to gaze on, burst out upon me. Nothing holds me ; I will indulge in my sacred fury. If you forgive me, I rejoice ; if you are angry, I can bear it. The die is cast, the book is written,—to be read either now or by posterity, I care not which : *it may well wait a century for a reader, as God has waited six thousand years for an interpreter of his works.*"

Sir John Herschel remarks of this law, " Of all the laws to which induction from pure observation has ever conducted man, this third law of Kepler may justly be regarded as the most remarkable, and the most pregnant with important consequences."

c. **Demonstration cf Kepler's Laws.**—These laws were deduced by Kepler, as matters of fact, from the recorded observations of himself and others ; but he failed to show the principle on which they are founded, and by which they are connected with each other. This was reserved for Newton, who, by the discovery and application of the law of gravitation, confirmed the truth of these laws by exact mathematical reasoning and calculation.

d. **Kepler's Third Law not quite true.**—The third law is, however, absolutely correct only when we consider the planets as mathematical points, without mass. Owing to the immense mass of the sun, this is relatively so nearly the fact, that the variation from the truth is very slight.

40. The eccentricity of the large planets' orbits is very small, that of Mercury being the greatest, and Venus the least. The orbits of the Minor Planets are generally remarkable for their great eccentricity.

a. **Comparative Eccentricities.**—The eccentricity of a planet's orbit is measured by comparing it with one-half of the major axis. The following is an approximate statement of the eccentricities of the large planets :—Mercury, $\frac{1}{5}$; Mars, $\frac{1}{10}$; Saturn, $\frac{1}{18}$; Jupiter, $\frac{1}{21}$; Uranus, $\frac{1}{22}$; Earth, $\frac{1}{60}$; Neptune $\frac{1}{117}$; Venus, $\frac{1}{148}$. The greatest of any of the minor planets is a little over $\frac{1}{3}$.

The annexed diagram will aid in giving the student a correct idea of the figure of the planets' orbits. This diagram represents an ellipse, the eccentricity of which is ⅓, or much greater than that of the most eccentric of the minor planets. It will be apparent, therefore, that the *actual figure* of the planets' orbits is but slightly different from that of a circle. If drawn on paper, the eye could not detect the difference.

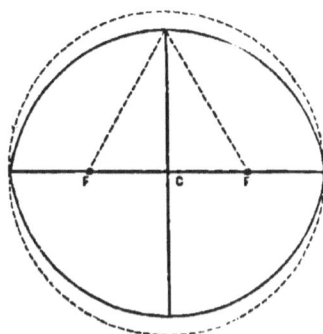

Fig. 23.

ELLIPSE—ECCENTRICITY, ⅓.

41. The MEAN PLACE of a planet is that in which it would be if it moved in a circle, and of course, with uniform velocity; the TRUE PLACE is that in which it is actually situated at any particular time.

42. The angular distance of the true place from the mean place, measured from the sun as a centre, is called the *Equation of the Centre.*

Fig. 24.

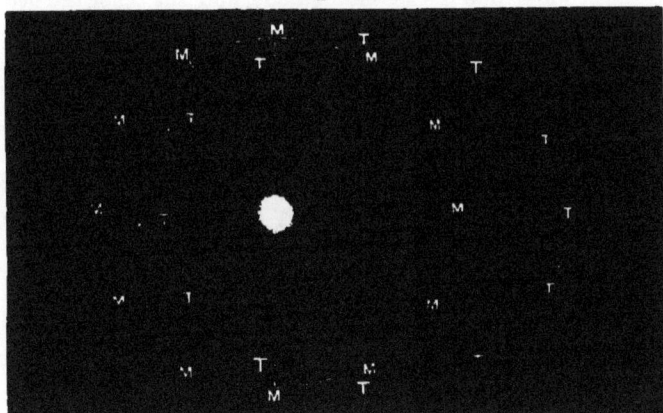

MEAN AND TRUE PLACES OF A PLANET.

In the above diagram, the ellipse represents the actual orbit of the planet, and the dotted circle the corresponding circular orbit. The points marked T represent the true places, and those marked M, the mean places of the planet. As the radius-vector passes over greater portions of the orbit in the perihelion than in the aphelion, the mean place is before or east of the true place, as the body moves from aphelion to perihelion, and behind or west of it in the other half of its revolution. The angle contained between the radius-vector and the radius of the circle is the equation of the centre.

43. The planets do not all revolve around the sun in the same plane, but in planes slightly inclined to each other. The angle which the plane of a planet's orbit makes with that of the earth's orbit is called the *Inclination of its Orbit.*

44. Of all the primary planets, Mercury has the greatest inclination of orbit (7°), and Uranus the least (46'). The Minor Planets are remarkable for a much greater inclination of their orbits than that of the other planets.

a. Since the planets' orbits are all inclined to that of the earth, each one must cross the plane of it in two points. These two points are called the **Nodes**; one the ascending node, and the other the descending node.

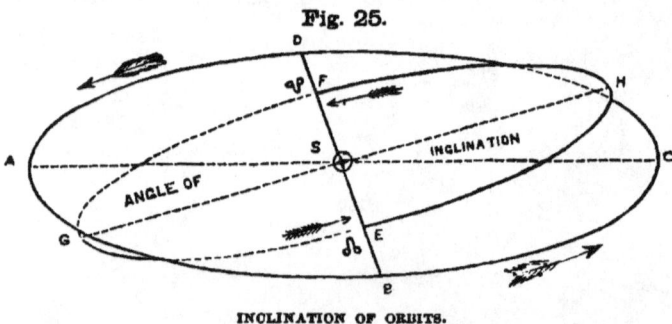

Fig. 25.

INCLINATION OF ORBITS.

Fig. 25 represents an oblique view of the orbits of the earth and Venus. E is the ascending, and F, the descending node. E F the line of nodes, and A S G the angle of inclination of the orbit.

45. The Nodes* of a planet's orbit are the two opposite points at which it crosses the plane of the earth's orbit.

46. The Ascending Node is that at which the planet crosses from south to north; the Descending Node, that at which it crosses from north to south. The straight line which joins these points is called the *Line of Nodes*.

☊ is the sign of the ascending node; ☋, of the descending node.

Fig. 26.

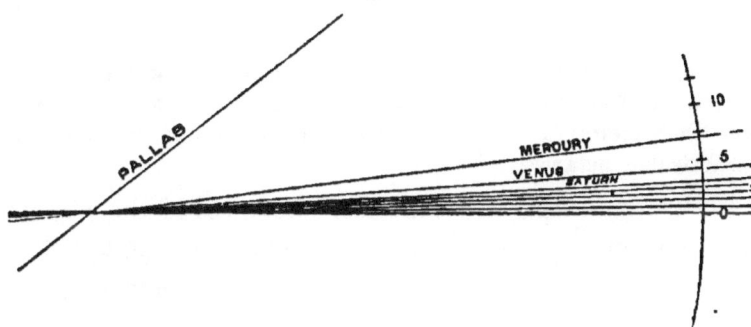

INCLINATION OF PLANETS' ORBITS.

Fig. 26 represents the position of the plane of each orbit in relation to that of the earth. The small amount of deviation from one uniform plane will be at once apparent. These planets, however, on account of their vast distance from the sun, depart very far from the plane of the earth's orbit. Thus, Mars, although having only 2° of inclination, may be nearly 5 millions of miles from this plane; and Neptune, about 85 millions.

* From the Latin word *nodus*, meaning *a knot*.

QUESTIONS.—45. What are nodes? 46. What is the ascending node? Descending node? Line of nodes?

CHAPTER V.

DISTANCES, PERIODIC TIMES, AND ROTATIONS OF THE PLANETS.

47. The distances of the planets from the sun are so great that they can only be expressed in millions of miles.

a. **Idea of a Million.**—A million is so vast a number that we can form no true conception of it without dividing it into portions. To count a million, at the rate of 5 per second, would require about 2½ days, counting without intermission, night and day. A railroad car, traveling at the rate of 30 miles per hour, night and day, would require nearly four years to pass over a million of miles. In stating the distances of the planets, the *rate of the express train* may be employed as a standard of comparison, so that the pupil may obtain something more than merely a knowledge of figures in learning these almost inconceivable distances.

48. The following are the mean distances of the planets from the sun, expressed in approximate round numbers :—

Mercury, .	35 millions.	Jupiter, .	476 millions.
Venus, .	66 "	Saturn, .	872 "
Earth, . .	91½ "	Uranus, .	1,754 "
Mars, . .	139 "	Neptune,	2,746 "

Minor Planets, . . 260 millions (average).

a. **Illustration.**—Multiply each of these numbers expressing millions by four, and we shall find the time which an express train starting from the sun would require to reach each of the planets. In the case of the nearest planet, this period would be 140 years, and of the most remote, almost 11,000 years. A cannon ball moving at the rate of 500 miles an hour, would not reach Neptune in less than 626 years.

---

b. **Bodé's Law.**—A comparison of the distances given above will show a very curious numerical relation existing among them, each distance being nearly double that next inferior to it. A more exact statement of this numerical relation was published in 1772 by Professor Bodé, of Berlin, although not discovered by him : it has usually been designated " Bodé's Law." Take the numbers

0,   3,   6,   12,   24,   48,   96,   192,   384 ;

each of which, excepting the second, is double the next preceding ; add to each 4, and we obtain

4,   7,   10,   16,   28,   52,   100,   196,   388 ;

which numbers very nearly represent the relative proportion of the planets' distances, including the average distance of the Minor Planets. In the case of Neptune, the law very decidedly fails, and, consequently, has ceased to have the importance attributed to it previous to the discovery of this planet in 1846.

PERIODIC TIMES OF THE PLANETS.

49. The following are the periods of time occupied by the planets respectively in completing one revolution around the sun :—

Mercury, . 88 days.   Jupiter, . 12 yrs. (nearly.)
Venus, . 224½ "   Saturn, . 29½ "
Earth, . . 365¼ "   Uranus, . 84 "
Mars, . . 1 yr. 322 days. Neptune, 165 "

Thus the year of Neptune is about 700 times as long as that of Mercury.

50. Of all the primary planets, Mercury moves in its orbit with the greatest velocity, and Neptune with the least ; the velocities of the planets diminishing as their distances from the sun increase.

a. This is in accordance with Kepler's third law ; since the ratio of the periodic times increases faster than that of the distances ; the square

QUESTIONS.—*b.* What is Bodé's law ? 49. State the periodic times of the primary planets. 50. Which planet moves with the greatest velocity ? Which, the least ? *a.* Why is this ?

of the former being equal to the cube of the latter. Thus, if the dis
tance of one planet is four times as great as that of another, the
periodic time will not be simply *four* times as long, but *eight* times as
long ; that is, the square root of the cube. ($\sqrt{4^3} = \sqrt{64} = 8$). Hence,
as the planet has a longer time in proportion to the distance traveled,
its velocity must be diminished.

b. **Comparative Velocities.**—The following table exhibits the mean
hourly motion of the primary planets in their orbits :—

Mercury, . . 104,000 miles.	Jupiter, . . 28,700 miles.
Venus, . . 77,000 "	Saturn, . . 21,000 "
Earth, . . . 65,500 "	Uranus, . . 15,000 "
Mars, . . . 53,000 "	Neptune, . . 12,000 "

c. **Illustration.**—What an amazing subject for contemplation does
this table present ! For example, the weight of the earth in tons is
computed to be about 6,000,000,000,000,000,000,000,000 ; that is to say, six
thousand million million times a million, or $6,000 \times 1,000,000 \times 1,000,$-
$000 \times 1,000,000$. Yet this body so inconceivably vast is rushing
through the abyss of space with a velocity of 1,000 miles per minute,
or about 15 miles during every pulsation of the heart. But the earth
in comparison with the body around which it is revolving is as a *single
grain* of wheat compared with *four bushels.*

d. **To find the Hourly Motion.**—This can be done by the application of very simple principles. The orbits being nearly circles, twice
the mean distance will give us the diameter, and 3¼ times the diameter
will give the circumference, or whole distance traveled in the periodic
time. Then finding the number of hours in this time, and dividing the
whole distance by this number, we obtain the hourly motion. Thus,
Mercury's mean distance is 35 million miles ; then $35 \times 2 \times 3\frac{1}{7} = 220$
millions, the whole distance traveled in 88 days, or $88 \times 24 = 2112$ hours ;
and 220 million ÷ 2112 = 104,166 miles.

AXIAL ROTATIONS OF THE PLANETS.

51. Besides revolving around the sun, the planets revolve
upon their axes in the same direction as they revolve in
their orbits; that is, from west to east. (See Art. 18, *a.*)
This is called their DIURNAL ROTATION.

QUESTIONS.—*b.* State the comparative velocities of the planets.   *c.* Illustration ?
d. How is the hourly motion in the orbit found ? 51. What is meant by *diurnal
rotation ?*

52. The Axis of a planet is the imaginary straight line passing through its centre, on which we conceive it to rotate.

53. A planet must rotate with its axis either perpendicular or oblique to the plane of its orbit. The axes of the planets are all considerably oblique, excepting that of Jupiter, which is only 3° from the perpendicular; that of Venus is supposed to be 75°.

54. The angle which the axis of a planet makes with a perpendicular to its orbit, is called its INCLINATION OF AXIS.

Fig. 27.

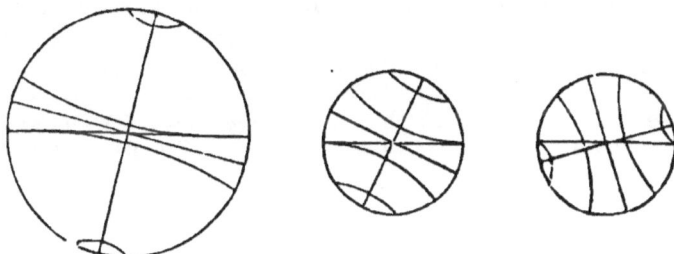

INCLINATION OF JUPITER, EARTH, AND VENUS.

a. The inclination of the axis of each planet, as far as it has been discovered, is as follows :—

Mercury, . . (unknown.)	Jupiter, . . 3°.
Venus, . . 75°. (?)	Saturn, . . 26¾°.
Earth, . . . 23½°.	Uranus, . . (unknown.)
Mars, . . . 28½°.	Neptune, . . "

b. **How to Discover the Rotation.**—The usual method of discovering the rotation of a planet is to examine the disc with a powerful telescope, so as to find, if possible, any spots upon it, and then to detect any regular movement of such spots across the disc. Let the pupil stand a short distance from a terrestrial globe, and let it be caused to revolve, and he will observe the marks upon it move across, and alternately disappear and re-appear. The same thing must, of course, occur in our observation of the planets, if they have a diurnal motion.

55. The times of rotation of the planets respectively are as follows:

Mercury, . . $24\frac{1}{4}$ hours.	Jupiter, . . 10 hours.
Venus, . . $23\frac{1}{2}$ "	Saturn, . . $10\frac{1}{2}$ "
Earth, . . . 24 "	Uranus, . . $9\frac{1}{2}$ " (?)
Mars, . . . $24\frac{1}{2}$ "	Neptune, . (unknown.)

a. It will be observed that the terrestrial planets all perform their rotations in about 24 hours; but that the major planets require less than one-half that time.

b. Sun's Rotation.—The sun also rotates upon an axis, but requires about 608 hours, or $25\frac{1}{3}$ days to complete one rotation. The inclination of its axis to the plane of the earth's orbit is about $7\frac{1}{3}°$.

QUESTIONS.—55. State the time of the rotation of each planet. *a.* What distinction, in this respect, between major and terrestrial planets? *b.* Does the sun rotate? In what time?

CHAPTER VI.

56. The ASPECTS of the planets are their apparent positions with respect to the sun or to each other. The principal aspects, that is, those most frequently referred to, are *Conjunction, Quadrature,* and *Opposition.*

57. A planet is said to be in CONJUNCTION with the sun when it is in the same part of the heavens.

That is, if the sun is in the east, the planet must also be in the east, both being seen, if visible, precisely in the same direction. It is evident that in the case of the inferior planets, this may occur in two ways; namely, when the planet is at that point of its orbit which is nearest to the earth or at the point most remote; or, in other words, when the earth and planet are both on the same side of the sun, or on opposite sides. Of course a superior planet, to be in conjunction, must be on the opposite side of the sun from the earth.

Fig. 28.

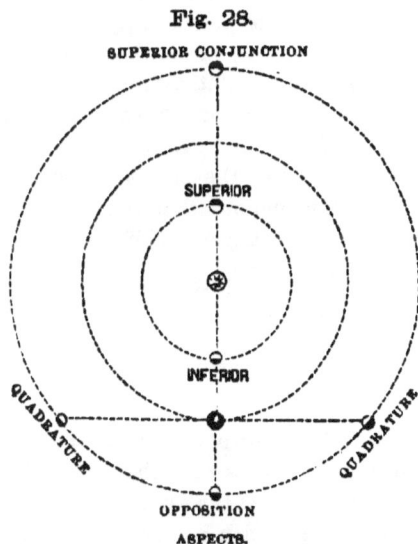

SUPERIOR CONJUNCTION

SUPERIOR

INFERIOR

QUADRATURE

QUADRATURE

OPPOSITION

ASPECTS.

58. Conjunction may be *Inferior* or *Superior.* Inferior

conjunction is that in which the planet is between the earth and the sun; superior conjunction is that in which the planet is on the opposite side of the sun from the earth.

59. A planet is said to be in OPPOSITION with the sun when it is in the opposite part of the heavens.

a. That is, while the sun is in the east, the planet, if in opposition, must be in the west. If Jupiter, for example, should be rising just as the sun is setting, or *vice versa*, it would be in opposition. It is obvious that the superior planets only can be in opposition, and that when in that position, they are at the points of their orbits nearest to the earth.

b. These different aspects obviously depend upon the angular, or apparent, distance of a planet from the sun. [See Introduction, Art. 18, *a*]. In conjunction, there is no angular distance, unless we regard the difference in the planes of the orbits; and when the planet is in conjunction and at either of the nodes, none whatever. In opposition, the angular distance is 180°.

60. The angular distance of a planet from the sun is called its ELONGATION.

61. A planet is said to be in QUADRATURE when its elongation is 90°.

a. The position of quadrature *in the heavens* is half-way between conjunction and opposition, the planet being so situated that the straight lines that connect the earth with the sun and planet, respectively, make a right angle with each other. Thus if a planet were in quadrature, it would be in the south, or near it, either at sunset or sunrise, according as it were either east or west of the sun. It will be obvious, from Fig. 28, that, viewed from the earth as a centre, the position of quadrature *in the orbit* is not half-way between conjunction and opposition, but much nearer the latter.

b. There are, in all, five aspects of the *planets*, depending on their relative positions. The following are their names, the angular distances, and the characters used to denote them :

Conjunction,	.	☌	0°.	Trine, . . .	△	120°.
Sextile, . . .		✳	60°.	Opposition, .	☍	180°.
Quartile, . . .		☐	90°.			

Fig. 29.

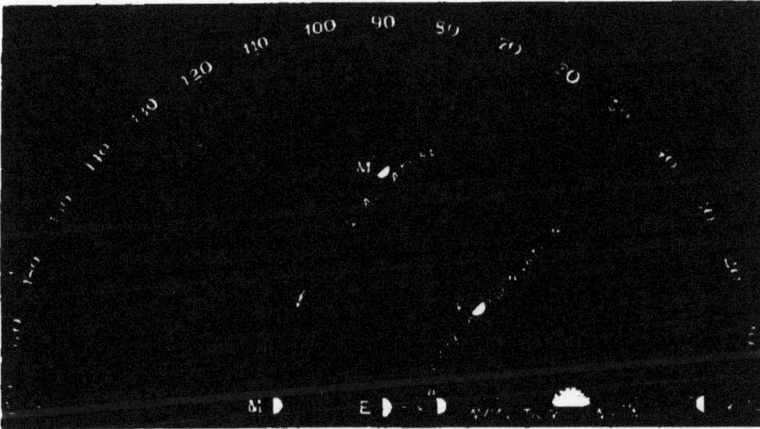

In the diagram, the graduated semicircle cuts the sides of all the angles which have their vertices at E, and serves to measure the angular distance of each planet from the sun. V and V″ represent Venus in superior and inferior conjunction, the elongation being, at those points, 0° ; while at V′, it is at its point of greatest elongation. It will be obvious from this diagram that no inferior planet can be 90° from the sun. M represents Mars in opposition, and M′ the same planet in quadrature. The aspect of M and V or V″ is opposition ; of M′ and V or V″, quartile.

62. The time which elapses between two similar elongations of a planet is called its *Synodic* * *Period.*

a. Thus the interval between two successive conjunctions or oppositions is the synodic period. The synodic period would be the true periodic time if the earth were at rest ; but the earth is moving in its orbit in the same direction as the planet, with a velocity less than that

---

* From the Greek words *syn*, meaning *together*, and *odos*, which means a *pathway.*

QUESTIONS.—62. What is the synodic period ? *a.* Why not the true period ? Illustrate by the diagram. How to calculate the synodic period of the inferior planets ? (Fig. 30.) Of the superior planets ? (Fig. 31.)

of the inferior planets and greater than that of the superior. Hence, the synodic period of an inferior planet must always be greater than the periodic time, while that of the superior planets is generally less.

Fig. 30.

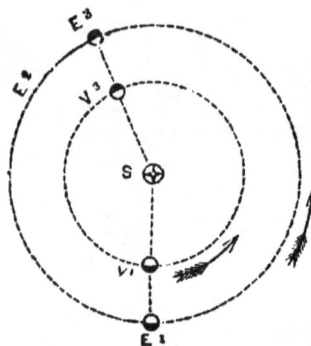

SYNODIC PERIOD. INFERIOR PLANETS.

The diagram represents Venus at V^1 in inferior conjunction with the sun, the earth being at E^1. Now conceive Venus to move around once, so as to return to V^1; the earth will then have gone over about 1⅗ of her orbit, and reached E^2, and Venus will not overtake her until she reaches E^3, passing her first position, and hence making *one* revolution, and the part E^1 E^3 besides, while Venus makes *two* revolutions, and of course a corresponding part of her orbit besides. This part of the orbit of each is about $\frac{1}{10}$ of the whole, in the case of Venus. For since Venus completes a revolution, or 360°, in 224¾ days, she moves about 1.6° per day; while the earth moves about .98° per day; hence Venus gains .62° per day; but she has 360° to gain, as she leaves V^1, and 360° ÷ .62° = 582 days. The true synodic period is 584 days. Now, 584 ÷ 224¾ = 2.6, number of revolutions of Venus during one synodic period; and 584 ÷ 365¼ = 1.6, number of revolutions of the earth; and 2.6 rev. — 2 rev. = $\frac{6}{10}$ rev. = E^1 E^3 or V^1 V^2.

Fig. 31.

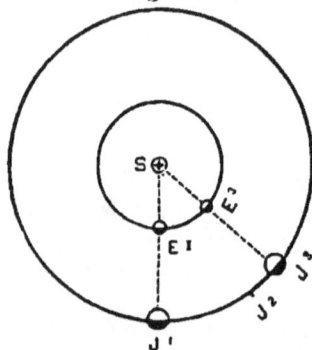

SYNODIC PERIOD, SUPERIOR PLANETS.

The synodic periods of the superior planets, are illustrated in the annexed diagram. Let J^1 represent Jupiter in opposition, the earth being at E^1. As Jupiter's periodic time is about 12 years, when the earth, after performing a revolution, returns to E^1, Jupiter has passed over $\frac{1}{12}$ of its orbit, and reached J^2, and the earth moving a short distance farther overtakes it at J^3. In this case, the superior planet only moves over a fraction of its orbit, while the earth moves over the same fraction of its orbit, and *one* whole revolution. We can find the synodic period of Jupiter from the true period, in the following manner:—As Jupiter performs only $\frac{1}{12}$ of a revolution while the earth performs a whole one, the earth gains $\frac{11}{12}$ of a revolution, while performing one; but to overtake Jupiter when starting from E^1, after opposition, she has to gain an entire revolution, and $1 \div 1\frac{1}{2} = 1\frac{1}{2}$. Now $1\frac{1}{2}$ of 365½ days = 399 days (nearly); which is the synodic period of Jupiter.

If the periodic time of any superior planet were exactly double that of the earth, its synodic period and periodic time would be equal. This is nearly true of Mars; its periodic time being 1 yr. 322 days, and its synodic period, 2 yrs. 50 days.

63. When a planet appears in the evening, just after sunset, it is called an *Evening Star ;* when in the morning, just before sunrise, it is called a *Morning Star.*

a. The inferior planets being always less than 90° from the sun, can only appear as morning or evening stars. Mercury being a small planet, and never having but a small amount of elongation, is a difficult object to see; Venus, being a large planet, and having a greater apparent distance from the sun, is a very brilliant and beautiful object, either as an evening or morning star. When the former, her elongation must of course be east; when the latter, west. The superior planets are morning or evening stars at different degrees of elongation, since they may be visible from sunset to sunrise.

Fig. 32.

VENUS AS MORNING AND EVENING STAR.

In the diagram, Fig. 32, Venus is represented as a morning and evening

QUESTIONS.—63. What is meant by morning star? By evening star? *a.* What is said of the inferior planets, in this respect? Explain from the diagram.

star. While Venus is on the side of the sun as represented at V, she must be an evening star, since, as the earth turns, any place at P must, as it turns from the sun at the time of sunset, still keep Venus in view by the angular distance contained between the lines drawn to the place from the sun and Venus respectively; but when Venus is at V', the other side of the sun, the rotation of the earth would bring Venus into view at any place as P, before the sun. (The student should carefully notice the direction of the motion as indicated by the arrows.) Venus, of course, remains the same side of the sun during one-half of the synodic period, or 292 days.

QUESTIONS FOR EXERCISE.

1. When a planet is in quadrature, what is its elongation?
2. What is its elongation when in inferior conjunction?
3. What is its elongation in superior conjunction?
4. How many degrees of elongation has it when in opposition?
5. Which of the planets can be in inferior conjunction?
6. Which can be in superior conjunction?
7. Which can be in opposition?
8. Which can be in quadrature?
9. Can the elongation of Mercury or Venus exceed 90°?
10. Can that of Jupiter?
11. What is the greatest elongation of a superior planet?
12. When Venus is in inferior conjunction, and Mars in opposition, what is their angular distance from each other? [See Fig. 29.]
13. What is their angular distance when Venus is in inferior conjunction, and Mars in superior conjunction?
14. How many degrees are they apart when Venus is in superior conjunction and Mars is in quadrature?
15. When the elongation of Venus is 30°, and that of Mars is 120°, what is their angular distance from each other?
16. If Venus is 50° from Mars, and the latter body is in quadrature, what is the elongation of Venus?

CHAPTER VII.

64. That the earth is, in its general form, a spherical body, is plainly indicated by a few simple facts :

1. Navigators are able to sail entirely around it either in an eastward or a westward direction ;

2. The earth and the sky always seem to meet in a circle, when the view is unobstructed ;

3. The top of a distant object always appears above this circle, before the lower parts ; as the sails of a ship before its hull ;

4. The elevation of the spectator causes this circle to sink, so as to show more of the earth's surface, and equally on all sides ;

5. The apparent movements of the heavenly bodies around the earth, some in large circles, some in small circles ; one particular star in the heavens not appearing to have any motion at all.

a. This last circumstance is accounted for by supposing that the earth's axis points to this star. Hence it is called the *North, or Pole Star.*

b. The first *practical proof* that the earth is spherical was afforded by the voyage of Magellan, whose squadron, in 1519–22, sailed entirely around the earth.

SECTION I.

LATITUDE AND LONGITUDE.

65. Points are located upon the surface of the earth by measuring their distances from certain established circles

---

QUESTIONS.—64. What five circumstances indicate that the general form of the earth is spherical? *a.* What is the north star? 65. How are points located on the earth's surface.

conceived to be drawn upon it. The position of these circles is determined by their relation to two fixed points,
called the POLES.

66. The poles are the two extremities of the earth's axis,
one being called the NORTH POLE, and the other the SOUTH
POLE.

a. As the earth turns on its axis from west to east, it causes all the
other heavenly bodies to seem to revolve around it from east to west,
in circles contracting in size towards the fixed point of the heavens,
called the *celestial pole*, near which is the pole-star.   The celestial
poles correspond to the poles of the earth, being the two points at which
the earth's axis, if extended, would meet the sphere of the heavens.

67. The great circle exactly midway between the two
poles is called the EQUATOR.   Its plane divides the earth
into northern and southern hemispheres.

68. The great circles that pass through the poles are
called *meridian circles ;* the half of a meridian circle that
extends from pole to pole, is called a *Me-
ridian.*

Fig. 33.

Meridians

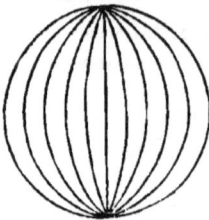

a. Meridian circles must, of course be perpendicular to the equator, and the plane of any
one of them would divide the earth into eastern
and western hemispheres. A great circle that
is perpendicular to another is sometimes called
a *secondary* to it.   Thus the meridian circles
are secondaries to the equator.

69. The position of a place on the surface of the earth is
indicated by its latitude and longitude. LATITUDE is distance north or south from the equator; LONGITUDE, distance
east or west from some established meridian, called a *First,*
or *Prime, Meridian.*

QUESTIONS —66. What are the poles ?   *a.* What are the celestial poles?   67. What is
the equator?   68. What are meridian circles?   Meridians?   *a.* Their relation to the
equator?   What is a secondary?   69. What is latitude?   Longitude?

Fig. 34.
Parallels

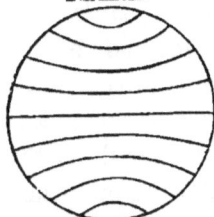

70. Small circles parallel to the equator are called PARALLELS OF LATITUDE.

71. Latitude is reckoned on a meridian, from the equator to the poles; longitude is reckoned from the prime meridian round to the opposite meridian.

a. Distance from any great circle must be reckoned on a secondary to that circle. It will be easily perceived by the pupil that the poles have the greatest possible latitude—namely, 90° ; and that places situated under the meridian opposite the prime meridian, have the greatest longitude, or 180° east or west; also, that a place situated at the intersection of the prime meridian with the equator can have neither longitude nor latitude.

b. **Difference of Time.**—Difference of Longitude causes difference of time. Since the earth turns toward the east, any place east of another place, must have *later time,* because it is sooner carried, by the motion of the earth, under the sun ; and, as an entire rotation, or 360°, is performed in 24 hours, 15° of longitude must be equivalent to one hour of time. Thus, London is 74° east of New York ; and, consequently, when it is noon at New York, it is 5 o'clock in the afternoon at London, the sun having passed the meridian five hours earlier.

c. **Difference of Longitude may be converted into Difference of Time,** by multiplying the degrees and minutes by 4 ; the former of which will then be minutes of time ; and the latter, seconds. For since $\frac{1}{15}$ the number of degrees is equal to the number of hours, $\frac{60}{15}$, or 4 times, the degrees must be equal to the minutes ; and, for the same reason, 4 times the minutes of space must be equal to seconds of time.

d. **To convert Difference of Time into Difference of Longitude,** reduce the hours to minutes, and divide by 4. For since 15 times the hours are equal to the degrees, $\frac{1}{60}$ of 15, or $\frac{1}{4}$, the minutes must be equal to the degrees.

QUESTIONS.—70. What are parallels of latitude? 71. How are latitude and longitude reckoned? *a.* Where is the latitude greatest? The longitude? What point or place on the earth's surface has neither latitude nor longitude? *b.* How does difference of longitude cause difference of time? *c.* How to convert difference of longitude into difference of time? *d.* How to convert difference of time into difference of longitude.

PROBLEMS FOR THE GLOBE.

PROBLEM I.—To find the latitude and longitude of a place: Bring the given place to the graduated side of the brass meridian [the circle of brass that encompasses the globe], which is numbered from the equator to the poles: and the degree of the meridian, over the place will be the latitude; and the degree of the equator, under the meridian, east or west of the prime meridian, will be the longitude.

Verify the following by the globe:

	LAT.	LONG.		LAT.	LONG.
LONDON, .	. . 51½° N.;	0°.	C. GOOD HOPE,	34° S.;	18½° E.
PARIS,	. . . 49° N.;	2¼° E.	BERLIN, .	. 52½° N.;	13½° E.
WASHINGTON,	. 39° N.;	77° W.	MADRAS,	. . 13° N.;	80° E.
CINCINNATI,	. 39° N.;	84½° W.	SANTIAGO, .	. . 32½° S.;	70½° W.

PROBLEM II.—The latitude and longitude of a place being given, to find the place: Find the degree of longitude on the equator, bring it to the brass meridian, and under the given degree of latitude, on the meridian, will be the place required.

EXAMPLES.

1. What place is in lat. 30° N., and long. 90° W.? *Ans.* New Orleans
2. What place "   " 42½° N.,   " 71° W.? *Ans.* Boston.
3. What place "   " 40¾° N.,   " 74° W.? *Ans.* New York.

PROBLEM III.—To find the difference of latitude or lon gitude between any two places : Find the latitude or longitude of both places; if on the same side of the equator or meridian, subtract one from the other; if on different sides, add them; the result will be the answer required.

EXAMPLES.

Find the difference of latitude and longitude of
1. London and Naples.   *Ans.* Lat. 10½°, long. 14¼°.
2. New York and San Francisco. *Ans.* Lat. 3°, long. 58½°.
3. Stockholm and Rio Janeiro. *Ans.* Lat. 82°, long. 61°.

PROBLEM *IV.*—*To find all the places that have the same latitude as any given place :* Bring the given place to the brass meridian, and observe its latitude ; turn the globe round, and all places that pass under the same degree of the meridian will be those required.

EXAMPLES.

What places have the same, or nearly the same, latitude as
1. MADRID ?   *Ans.* Minorca, Naples, Constantinople, Kokand, Salt Lake City, Pittsburgh, New York.
2. HAVANA ?   *Ans.* Muscat, Calcutta, Canton, C. St. Lucas, Mazatlan.

PROBLEM *V.*—*To find the places that have the same longitude as any given place :* Bring the given place to the graduated side of the brass meridian, and all places under the meridian will be those required.

EXAMPLE

What places have the same, or nearly the same, longitude as
LONDON ?   *Ans.* Havre, Bordeaux, Valencia, Oran, Gulf of Guinea.

PROBLEM *VI.*—*A time and place being given, to find what o'clock it is at any other place :* Bring the place at which the time is given to the brass meridian, set the index to the given time, and turn the globe till the other place comes to the meridian, and the index will point to the time required.

NOTE.—If the place be east of the given place, turn the globe westward; if west, turn it eastward.

This problem can be performed without the globe by finding the difference of longitude, as indicated in Art. 71, *c, d.*

EXAMPLES.

1. When it is noon at New York, what o'clock is it at London ? *Ans.* 5 o'clock P.M. (nearly).
2. When it is 10 o'clock A.M. at St. Petersburg, what o'clock is it at the City of Mexico ? *Ans.* 1 hour 20 min. A.M.
3. When it is 9 o'clock P.M. at Rome, what o'clock is it at San Francisco ? *Ans.* Noon.

PROBLEM VII.—To find the distancce between any two places : Lay the graduated edge of the quadrant over both places, so that the division marked 0 may be on one of them; and the number of degrees between them, reduced to miles, will be the distance required.

NOTE.—If geographic miles are required, multiply the degrees by 60; if statute miles, by 69½.

EXAMPLES.

Find the distance in geographic and statute miles between
1. NORTH CAPE and CAPE MATAPAN. *Ans.* 2,100 geog. miles ; 2,413⅗
   statute miles.
2. RIO JANEIRO and CAPE FAREWELL. *Ans.* 4,980 geog. miles ; 5,736¼
   statute miles.

SECTION II.

THE HORIZON.

72. The HORIZON * OF A PLACE is the circle which separates the visible part of the heavens from the invisible.

a. The surface of the earth appears, to a person standing upon it, like a great plane, extending equally on all sides, and limited by a circle at which the earth and sky appear to meet. As the elevation of the spectator increases, the greater is the extent of surface embraced within this circle, and the more extensive the visible heavens as compared with the invisible. On the other hand, an eye situated exactly on the earth's surface sees but a point of it, but still beholds a circle bounding the visible heavens, the plane of which would touch the earth's surface at the exact point where the eye is located. This circle is called the *Sensible* or *Visible Horizon ;* and the depression of it, due to the elevation of the spectator, the *Dip of the Horizon.* The following definitions may therefore be given of each :

73. The SENSIBLE HORIZON is that circle of the celestial

---

* From the Greek word *horizo,* meaning *to bound.*

QUESTIONS.—72. What is the horizon of a place ? *a.* General phenomena connected with the horizon ? 73. What is the sensible horizon ?

sphere the plane of which touches the earth at the place of the spectator.

a. By the *Celestial Sphere* is meant the concave sphere of the heavens, in which the heavenly bodies appear to be placed, the observer being at the centre within, and looking upward.

74. The DIP OF THE HORIZON is the depression of the sensible horizon caused by the elevation of the spectator, and bringing a circular portion of the earth's surface into view.

In the diagram, let the small circle whose centre is E, represent the earth, the portion of the large circle V Z V' a part of the celestial sphere, and P the point, or place, of the spectator. Then the tangent S P S will represent the plane of the sensible horizon, and S Z S the visible heavens. Conceive the observer to stand above the surface at H; the tangents H V and H V' will then, at their points of contact, D and D, limit the visible part of the earth's surface, and at their extremities, V and V', the visible heavens. S V or S V' will be, of course, the dip of the horizon. At the point P, the visible part of the heavens is less than the invisible; but at so great an elevation as H P (represented as about 1,000 miles), the visible part would be much greater than the invisible, and a large part of the earth's surface, denoted by the arc D D, would come into view. The dip, however, at any attainable height is very small, and only an inconsiderable portion of the earth's surface can ever be seen. The line R R represents the plane of a great circle, which divides the celestial sphere into equal parts, passing through the centre of the earth, and situated at a distance from the plane of the sensible horizon equal to the semi-diameter of the earth, or nearly 4,000 miles.

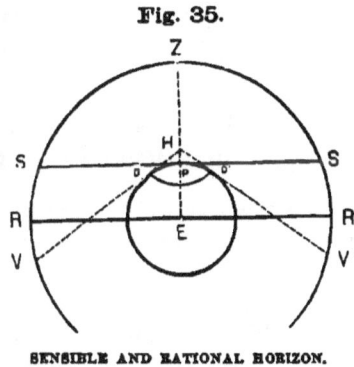

Fig. 35.

SENSIBLE AND RATIONAL HORIZON.

75. The great circle of the celestial sphere which is parallel to the sensible horizon, is called the RATIONAL HORIZON.

It divides the earth and the celestial sphere into upper and lower hemispheres.

The terms *upper* and *lower, above* and *below,* and the like, are only applicable to the horizon. The rational horizon is the real horizon : it is the standard circle for referring the *apparent* positions of all the heavenly bodies.

76. The poles of the horizon are called the ZENITH and the NADIR. The zenith is the point directly overhead ; the nadir is the point opposite to the zenith, and directly under our feet.

The one is the pole of the visible, or upper, hemisphere ; the other, the pole of the invisible, or lower. Each is, of course, 90° from the horizon.

77. Great circles conceived to pass through the zenith and nadir are called VERTICAL CIRCLES, or VERTICALS.

a. Vertical circles, being perpendicular to the horizon, are secondaries to it. The position of a body in the celestial sphere is defined by its distance from the *rational horizon,* and some selected vertical circle ; just as the position of a place on the earth's surface is determined by its latitude and longitude. The vertical selected for this purpose is that which the centre of the sun reaches and passes at noon. This circle, of course, passes through the north and south points of the horizon, and also through the celestial poles, its plane intersecting the earth so as to form a terrestrial meridian. It is therefore called the *Meridian of the Place.*

78. The MERIDIAN OF A PLACE is the vertical circle which passes through the north and south points of the horizon of that place. It divides the celestial sphere into eastern and western hemispheres.

a. When a body is on the meridian, it is said to *culminate,* because it is at that time at its greatest distance above the horizon during 24 hours.

QUESTIONS.—76. What are the zenith and the nadir ? 77. What are vertical circles ? *a.* How is the position of a body in the celestial sphere defined ? 78. What is the meridian of a place ? *a.* When is a body said to culminate ?

79. The distance of a body above the horizon is called its ALTITUDE. It is reckoned on a vertical circle, from the horizon to the zenith.

At the horizon, therefore, the altitude is 0° ; at the zenith, 90°.

80. The distance of a body east or west from the meridian is called its AZIMUTH. It is reckoned on the horizon.

a. **Prime Vertical; Amplitude.**—The altitude and azimuth of a body would be sufficient to define its position in the visible heavens ; but astronomers sometimes employ another vertical as a standard of reference, namely, that which passes through the east and west points of the horizon, cutting the meridian at right angles. This is called the *Prime Vertical ;* and the distance of a body from it, north or south, is call the *Amplitude.* These are, at present, but little used. By the amplitude of the sun is generally meant the distance at which it rises from the east, or sets from the west point of the horizon.

In the diagram, let N E S W represent the rational horizon, the circle passing through N S, the meridian, and that passing through E W, the prime vertical; then if A be the position of the sun at rising, A E will represent its amplitude, and A N, its azimuth; the altitude being 0c.

Fig. 36.

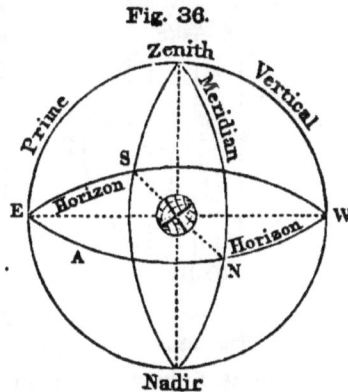

81. The ZENITH DISTANCE of a body is its distance from the zenith reckoned on a vertical circle.

The zenith distance is the *complement* of the altitude, that is, the difference between it and 90°.

82. The circles which the heavenly bodies may be conceived to describe during their apparent daily revolution around the earth, are called CIRCLES OF DAILY MOTION.

Fig. 37.

Parallel Sphere

Fig. 38.

Right Sphere

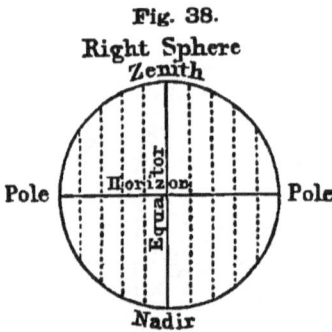

Fig. 39.

Oblique Sphere

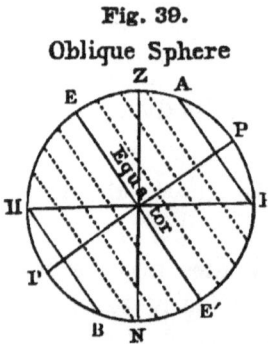

a. **Positions of the Sphere.**—The circles of daily motion are parallel, perpendicular, or oblique to the horizon, according to the place of the observer upon the surface of the earth. When standing exactly at either of the poles, he would have the celestial pole in the zenith, and the circles of daily motion would be parallel to the horizon; this position is called a *Parallel Sphere.* At the equator, the celestial poles would be in the horizon, and the circles of daily motion perpendicular to it; this position is called a *Right Sphere.* At any place between the equator and the pole the circles would be oblique to the horizon, and the pole would be raised to an altitude equal to the latitude of the place; this is called an *Oblique Sphere.*

b. In a parallel sphere, one-half of all the circles of daily motion are wholly above the horizon, and the heavenly bodies do not appear to rise and set, but to move around in parallel circles contracting in size toward the zenith; in a right sphere, all the circles are divided equally by the horizon, there being as much of each above as below it; in an oblique sphere, some of the circles of daily motion are wholly above the horizon, others wholly below it, and all between these, divided unequally by it. All this will be rendered apparent by the accompanying diagrams.

83. The circle of an oblique sphere in which the stars never set is called the CIRCLE OF PERPETUAL APPARI-

---

QUESTIONS.—*a.* What are the three positions of the sphere? Define each. *b.* How are the circles of daily motion divided by the horizon in each? 83. What is the circle of perpetual apparition? Of perpetual occultation?

TION; that in which they never rise, the CIRCLE OF PERPETUAL OCCULTATION.

84. That part of a circle of daily motion which is above the horizon, and which a body describes from its rising to its setting, is called the DIURNAL ARC; the part below the horizon is called the NOCTURNAL ARC.

In the diagram of the oblique sphere (Fig. 39), H H represents the rational horizon, Z and N the zenith and nadir, P P the poles, E E'the equator extended to the heavens, and the dotted lines, circles of daily motion. Then Z E will be the same number of degrees as the latitude, E H will be the altitude at which the equinoctial or equator intersects the meridian, and P H will be the altitude of the celestial pole. Now E P is equal to Z H, each being 90°; hence, by subtracting Z P from each, we find E Z = P H; that is, *the altitude of the pole equal to the latitude.*

PARALLAX.

85. The TRUE ALTITUDE of a body is the distance at which it would appear to be from the horizon, if it could be viewed from the centre of the earth.

In the diagram let the small circle represent the earth, having its centre at E; A, B, and C, a body as seen at different altitudes from the place, P; E H, the plane of the rational horizon: P h, the plane of the sensible horizon, and E Z, the direction of the zenith. At A, the body being in the sensible horizon, its apparent altitude will be nothing; but if viewed from E, it would appear to be above the horizon a distance equal to the angle m E H, or its equal m A h,

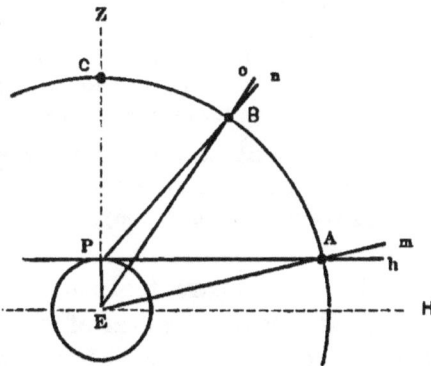

Fig. 40.

since the difference in direction between the lines E H or P h, and E m is the difference between the apparent and true altitude. At B, there is evidently a less difference of direction between the lines P n and E o, and when the body is at C, the centre of the earth, the place of the observer, and the position of the body being all on the same straight line, the true is the same as the apparent altitude. It is evident that the apparent altitude is always less than the true altitude, except when the body is seen in the zenith, as at C; and that there is the greatest difference when the body is in the horizon, as at A.

86. The difference between the true and apparent altitude of a heavenly body is called its PARALLAX.

87. The parallax of a body is greatest when it is in the horizon, and diminishes towards the zenith, where it is nothing. The parallax of a body when in the horizon is called its *Horizontal Parallax.*

In the preceding diagram, the angle m A h, or its equal P A E, is called the *angle of parallax.* o B n, or P B E, is the angle of parallax for the position B. The *angular* distance of the sensible and rational horizons is, of course, the horizontal parallax.

a. *The greater the distance of a body from the earth, the smaller is the angle of parallax.*

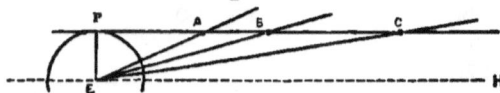
Fig. 41.

Thus the horizontal parallax of a body at A, (Fig. 41.) is A E H, or P A E; but at B, it is the smaller angle B E H, or P B E. The horizontal parallax of any body is really the angle subtended by the semi-diameter of the earth at the distance of the body; and, of course, the greater the distance, the smaller the angle.

b. The horizontal parallax of the moon is nearly 1°; that of the sun, less than 9″. In a subsequent chapter, it will be shown that by finding the parallax of a body, we can determine its distance from the earth.

88. Since the apparent altitude of a body is less than the true altitude by the amount of parallax, the effect of parallax is said to be to diminish the altitude. The apparent

altitude is therefore corrected by adding the amount of parallax due to the particular elevation and the distance of the body.

a. Other corrections would also have to be made to obtain the exact altitude ; namely, for the dip of the horizon caused by the elevation of the spectator, and for the effect of the atmosphere upon the direction of the rays of light which pass through it. The latter of these is called *Refraction.*

REFRACTION.

89. REFRACTION, in astronomy, is the change of direction which the rays of light undergo in passing through the earth's atmosphere.

a. It is a general fact that the rays of light when passing *obliquely*, from one medium into another of a different density, are turned from their course, and made to pass more obliquely, if the medium which they enter is rarer, and less obliquely, if it is denser than that which they leave. Thus, in passing from air into water, or from water into glass, the direction would be less oblique ; but in passing from water into air, more oblique.

Suppose *n m* to represent the surface of water, and S O a ray of light, entering the water at O. Instead of keeping on in the direction S O, it is bent toward the perpendicular A B, and thus passes less obliquely.

Fig 42.

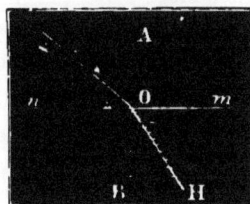

b. Now, as the earth's atmosphere is not of uniform density, but grows more and more dense toward the surface of the earth, the rays of light which proceed from any body are constantly bent more and more toward a perpendicular direction ; and since we see an object in the direction in which the ray of light strikes the eye, the apparent altitude of the body will be increased.

Fig. 43.

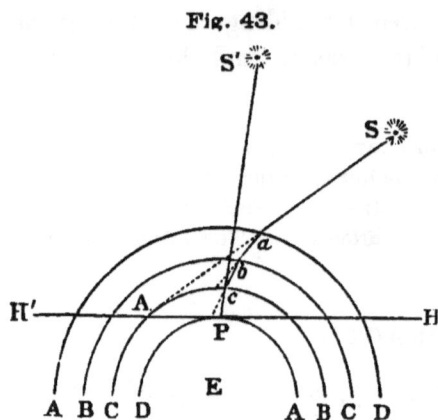

Suppose E to represent the earth, and A B C D, portions or strata of the atmosphere, of different densities, P, the place of observation. Suppose a ray of light from the star S, strike the atmosphere at a; on account of refraction, instead of proceeding in the direction S A, it describes a b, b c, and c P, reaching the spectator at P, and in the direction of c P; so that the star appears in that direction at S', and is thus elevated above its true position at S. As the atmosphere does not consist of distinct strata, as represented, but diminishes uniformly in density from the surface of the earth, the broken line a b c P, is in reality a curve, and the line S' P, a tangent to it at the point P.

90. The effect of refraction is greatest upon a body when it is in the horizon, and diminishes toward the zenith, where it is nothing. At the horizon, it amounts to about 33 minutes.

a. There is no refraction at the zenith, because at that point every ray of light strikes the atmosphere perpendicularly, and refraction only takes place when the direction of the rays is oblique; at the horizon, they are more oblique than they can be at any point above it; hence the refraction is greatest there.

91. At the horizon, the amount of refraction is somewhat greater than the apparent diameter of the sun or moon; and hence these bodies appear to be above the horizon when they are actually below it.

a. The times of the rising of all the heavenly bodies are, therefore, accelerated, and those of their setting retarded, by refraction; each one appears to be above the horizon before it has actually risen, and is seen above the horizon after it has actually set.

QUESTIONS.—90. What is the effect of refraction at the zenith and horizon? Why? 91. What is the amount of refraction at the horizon? *a.* Effect on the rising and setting of the heavenly bodies?

b. Refraction very rapidly diminishes from the horizon towards the zenith. At the horizon its mean value is 33′; at 10° of altitude, 15¼′; at 30°, 1½′; at 45°, 57″; at 80°, 10″; at 90°, 0.

SECTION III.

APPARENT MOTIONS OF THE SUN AND STARS.

92. The sun has two apparent motions around the earth; namely, a diurnal motion from east to west, and an annual motion from west to east. The first is caused by the rotation of the earth on its axis, and the second, by its revolution around the sun.

a. The student should be careful to verify by his own observations the following statements respecting the sun's apparent motions :—

1. **Apparent Daily Motion.**—The sun rises exactly at the east point of the horizon, and sets at the west point, twice a year; namely, about the 20th of March and 23d of September; and, on these days, it crosses the meridian at an altitude equal to the complement of the latitude; that is, at the point where the celestial equator crosses the meridian.

2. From March 20th to June 21st, the points at which the sun rises and sets move from the east and west toward the north, and its meridian altitude constantly increases; from June 21st till Sept. 23rd, the points of rising and setting move back toward the east and west, and the meridian altitude diminishes; from Sept. 23rd to Dec. 22nd, the points of rising and setting move toward the south, and the meridian altitude diminishes; from Dec. 22nd to March 20th, the points of rising and setting move back toward the east and west, and the meridian altitude increases. There is thus a constant movement of the points of rising and setting alternately from north to south, and a constant variation, up and down, of the point of culmination, except that the sun culminates at the same altitude for several days, about the 21st of

Questions.—*b.* How fast does refraction diminish from the horizon? 92. What apparent motions has the sun? How caused? *a.* State the daily phenomena connected with the apparent motions of the sun. What changes in the points of rising and setting? In the point of culmination? Solstices and equinoxes?

June and the 22nd of December. These two stationary points of culmination are called the *Solstices*.* The points at which the culmination of the sun coincides with that of the celestial equator are called
the *Equinoxes*,† because when the sun is at either of these points, the
days and nights are exactly equal to each other.

3. **Apparent Annual Motion.**—The sun appears to move toward
the east among the stars ; for, if on any evening at sunset, or a short
time after, we notice the distance of the sun from any star that may be
visible, we shall find, in a few evenings, that this distance has grown
less ; and hence, as the stars are fixed points, that the sun has moved
toward the east. This motion will continue from month to month
until the sun will be in conjunction with the star ; and then for six
months the star will be no longer visible, but at the end of that time,
will show itself above the eastern edge of the horizon just as the sun
sets below the western ; and at the expiration of one year from the
first observation, will have returned to the same relative position with
the sun. In this way the sun appears to move from star to star toward
the east, completing its circuit in 365¼ days.

93. The great circle of the celestial sphere in which the
sun appears to revolve around the earth every year, is called
the ECLIPTIC.

The ecliptic may also be defined as the great circle of the celestial
sphere in which it is intersected by the plane of the earth's orbit. Hence
the plane of the ecliptic is the plane of the earth's orbit.

94. The great circle of the celestial sphere exactly over
the equator is called the EQUINOCTIAL, or CELESTIAL
EQUATOR.

The student must conceive these circles as marked out on the sky,
the one crossing the other. (See diagram, Fig. 44.)

95. Since the earth's axis is inclined to the plane of its

---

* From the Latin words *sol*, meaning the *sun*, and *sto*, meaning *to stand*.

† From the Latin words *equus*, meaning *equal*, and *nox*, meaning *night*. The
arrival of the sun at either of these points produces equal days and nights.

QUESTIONS.—State the phenomena connected with the sun's apparent annual motion.
93. What is the ecliptic? 94. What is the equinoctial? 95. What is meant by the
obliquity of the ecliptic? Why is it 23½°?

orbit, or the plane of the ecliptic, making with it an angle of $66\frac{1}{2}°$, the ecliptic and equinoctial must cross each other at an angle of $23\frac{1}{2}°$. This angle is called the OBLIQUITY OF THE ECLIPTIC.

Fig. 44.

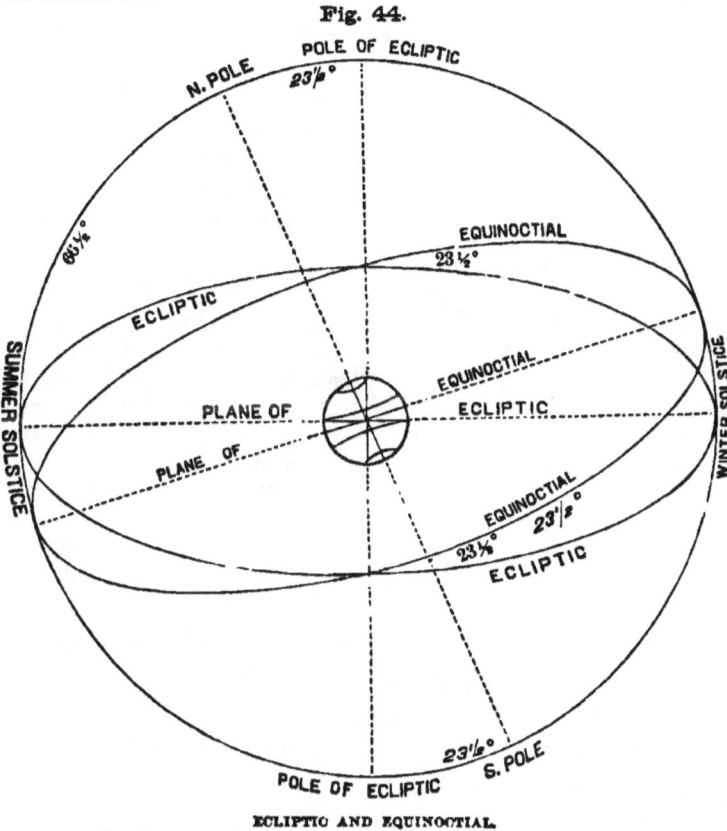

ECLIPTIC AND EQUINOCTIAL.

a. The obliquity of the ecliptic, and, of course, the inclination of the axis, are indicated by the difference between the highest and lowest daily culminating points of the sun, being equal to one-half of this dif- ference. For when it is at the equinoctial, it must culminate where the equinoctial crosses the meridian, that is, at an altitude equal to the complement of the latitude ; and when it is north or south of the equinoctial, it must culminate as far above or below the culminating point of the equinoctial. But this never exceeds 23½° either way; hence, the obliquity or inclination must be 23½°. This departure of the sun from the equinoctial, as indicated by its daily motion, is called its *Declination*.

b. *To find the greatest and least meridian altitude of the sun at any place*, the following rule may be given : Find the complement of the latitude, and to it add 23½° for the greatest altitude; and from it subtract 23½° for the least. Thus for New York the lat. of which is about 40½°: 90° − 40½° = 49½° comp. of lat. Hence, 49½° + 23½° = 73°, greatest altitude; and 49½° −23½° = 26°, least altitude.

Fig. 45.

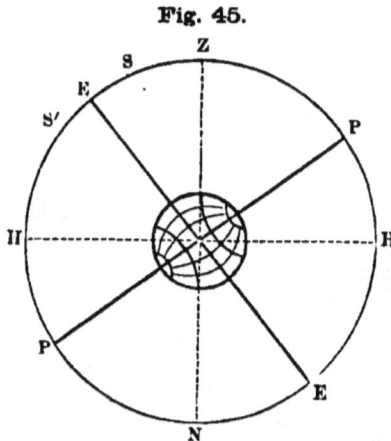

GREATEST AND LEAST ALTITUDE OF THE SUN.

In the diagram, P P represents the celestial poles, E E the equinoctial, Z N the zenith and nadir, H H the horizon, S the position of the sun when 23½° north of the equinoctial, and S′ its position when 23½° south of the equinoctial; then E H will represent its altitude when at the equinoctial, E H + E S, its greatest meridian altitude; and E H − E S′, its least.

96. The two opposite points of the ecliptic, where it crosses the equinoctial, are called the EQUINOCTIAL POINTS, or EQUINOXES. The one which the sun passes in March is called the *Vernal Equinox ;* that which it passes in September, the *Autumnal Equinox.*

97. The two opposite points of the ecliptic at which the sun is farthest from the equinoctial, are called the SOLSTITIAL POINTS, or SOLSTICES. The one north of the equinoctial is called the *Summer Solstice ;* the one south of it, the *Winter Solstice.*

QUESTIONS.—*b.* How to find the greatest and least meridian altitude of the sun? Explain by the diagram. 96. What are the equinoxes? How distinguished? 97. The solstices, and how distinguished?

a. The equinoxes and solstices are sometimes called the cardinal points of the ecliptic ; they are 90° from each other, and, of course, divide the ecliptic into four equal parts.

98. The DECLINATION of a heavenly body is its distance, north or south, from the equinoctial.

a. Declination corresponds to terrestrial latitude. At the equinoxes, the declination of the sun is 0° ; at the solstices, it is 23½°, which is the greatest declination the sun can have.

b. In a right sphere, the amplitude of the sun when it is rising or setting, is exactly equal to its declination. [Let the student verify this by an artificial globe.]

99. CIRCLES OF DECLINATION are great circles of the celestial sphere that pass through the poles, and are perpendicular to the equinoctial.

a. **Hour Circles.**—Circles of declination correspond to meridian circles on the earth. When drawn at intervals of 15°, they are called *Hour Circles*, because the heavenly bodies, in their apparent diurnal revolution round the earth, pass from one to the other every hour ; since 360° ÷ 24 = 15°.

b. **Hour Angle.**—The angle included between the hour circle passing through a body and the meridian of the place of observation is called the *Hour Angle* of the body.

c. **Colures.**—The circle of declination that passes through the equinoctial points is called the *Equinoctial Colure ;* that which passes through the solstitial points is called the *Solstitial Colure.*

d. The position of a heavenly body in the celestial sphere is defined by its distance from the equinoctial (or declination), and its distance from the equinoctial colure, reckoned eastward from 0° to 360°. The latter is called its *Right Ascension.*

100. The RIGHT ASCENSION of a heavenly body is its distance from the equinoctial colure, reckoned on the equi-

QUESTIONS.—*a.* What are these points sometimes called ? 98. What is declination ? *a.* Greatest declination of the sun ? *b.* Amplitude of the sun in a right sphere—equal to what ? 99. What are circles of declination ? *a.* What are hour circles ? *b.* What is the hour angle ? *c.* What are colures ? *d.* How is the position of a body in the celestial sphere defined ? 100. What is right ascension ?

noctial from the vernal equinox eastward entirely around the circle, that is, from 0° to 360°.

a. Right ascension is frequently expressed in hours, minutes, and seconds ; reckoning, of course, 15° to an hour. (See Art. 71, *b.*) Thus, 150° 30′ 15″ = 10ʰ 2ᵐ 1ˢ.

SIGNS OF THE ECLIPTIC.

101. The ecliptic is divided into twelve equal parts, called SIGNS. Each sign, therefore contains 30 degrees.

102. The following are the names of the signs, the characters denoting them, and the day of the month on which the sun enters each (1867):

Spring Signs.	ARIES	♈	March 20.	VERNAL EQUINOX.
	TAURUS	♉	April 20.	
	GEMINI	♊	May 21.	
Summer Signs.	CANCER	♋	June 21.	SUMMER SOLSTICE.
	LEO	♌	July 23.	
	VIRGO	♍	August 23.	
Autumn Signs.	LIBRA	♎	September 23.	AUTUMNAL EQUINOX.
	SCORPIO	♏	October 23.	
	SAGITTARIUS	♐	November 23.	
Winter Signs.	CAPRICORNUS	♑	December 22.	WINTER SOLSTICE.
	AQUARIUS	♒	January 20.	
	PISCES	♓	February 18.	

a. The equinoctial points are, it will be observed, at the first degree of Aries and Libra ; and the solstitial points at the first degree of Cancer and Capricorn.

103. The ZODIAC is a zone of the celestial sphere, extending to the distance of eight degrees on each side of the ecliptic.

a. The zodiac is therefore 16° wide ; and within its limits are con-

---

QUESTIONS.—*a.* How is right ascension frequently expressed? 101. How is the ecliptic divided? 102. Name the signs, and the day on which the sun enters each. *a. At which of the signs are the equinoxes and solstices ?* Date of each ? 103. What is the zodiac? *a.* What is its width? What does it contain?

tained the orbits of all the planets, except some of the Minor Planets. It also contains twelve of the great groups of stars, called the *Constellations of the Zodiac*, which have the same names, and occupy nearly the same places, as the signs of the ecliptic. The names given to the signs are properly the names of the constellations : Aries, *the ram ;* Taurus, *the bull ;* Gemini, *the twins ;* Cancer, *the crab ;* Leo, *the lion ;* Virgo, *the virgin ;* Libra, *the balance ;* Scorpio, *the scorpion ;* Sagittarius, *the archer ;* Capricornus, *the goat ;* Aquarius, *the water-carrier ;* Pisces, *the fishes.*

b. The places of the signs nearly corresponded with those of the constellations in the time of Hipparchus, by whom the constellations of the sphere were classified and arranged. He was the first to discover (125 B. C.) the displacement of the signs, and to explain its cause,—the falling back (from east to west) of the equinoxes. This movement of the equinoctial points is called *precession.*

PRECESSION.

104. PRECESSION is a gradual falling back of the equinoctial points from east to west.

In other words, the sun, in his apparent annual revolution around the earth, does not cross the equinoctial always at the same points, but at every revolution crosses a little to the west of the points at which it crossed previously. The irregularity seems to exist in the motion of the sun ; but, of course, it is really in the motion of the earth.

105. The amount of precession annually is about 50 seconds (50.2″); and consequently, to pass quite round the circle, the equinoxes require a period of nearly 26,000 years.

a. For $360° \times 60 \times 60 = 1,296,000″ \div 50.2″ = 25,816.$

b. In the time of Hipparchus, the vernal equinox was in the constellation Aries ; but it is now in Pisces, having fallen back about 28°. The signs and constellations corresponded about 185 B. C.

c. In maps of the heavens, and catalogues of the stars, the places of stars are marked by their declinations and right ascensions ; in some, however, they are indicated by their *latitudes* and *longitudes,* which terms, when applied to celestial objects, have a different meaning from that which they have as applied to places on the earth's surface.

CELESTIAL LATITUDE AND LONGITUDE.

106. The LATITUDE OF A HEAVENLY BODY is its distance from the ecliptic, north or south.

Celestial latitude must of course be reckoned upon a secondary to the ecliptic, from 0° to 90°.

107. The LONGITUDE OF A HEAVENLY BODY is its distance from a secondary to the ecliptic which passes through the vernal equinox, or first degree of Aries. It is reckoned from the vernal equinox eastward from 0° to 360°.

a. Of course, neither the longitude nor right ascension of a body can be quite equal to 360° ; since that would bring it back to the point of commencement, or 0° ; it may, however, be any distance less than that ; as, 359° 59′ 59″.

PROBLEMS FOR THE TERRESTRIAL GLOBE.

PROBLEM I.—*To find the sun's longitude for any given day :* Look for the given day of the month on the wooden horizon, and the sign and degree corresponding to it, in the circle of signs, will be the sun's place in the ecliptic; find this place on the ecliptic, and the number of degrees between it and the first point of Aries, counting toward the east, will be the sun's longitude.

EXAMPLES.

1. What is the longitude of the sun June 21st ? *Ans.* 90°.
2. What is it February 22d ? *Ans.* 334½°.
3. What is it May 10th ? *Ans.* 50°.

PROBLEM II.—To find the right ascension of the sun: Bring the sun's place in the ecliptic to the edge of the brass meridian; and the degree of the equinoctial over it, reckoning from the first degree of Aries, toward the east, will be the right ascension.

EXAMPLES.

1. What is the right ascension of the sun October 18th ?   *Ans.* 203¼°.
2. What is it May 2d ?   *Ans.* 42°.

PROBLEM III.—To find the declination of the sun: Bring the sun's place in the ecliptic to the edge of the brass meridian; and the degree of the meridian over it, reckoning from the equator, will be the declination. The declination may also be found by bringing the given day of the month as marked on the *analemma* to the meridian.

EXAMPLES.

1. What is the declination of the sun June 21st ?   *Ans.* 23½° N.
2. What is its declination Jan. 27th ?   *Ans.* 18¼° S.
3. What is it April 16th ?   *Ans.* 10° N.

PROBLEM IV.—To find what places have a vertical sun on any day in the year : Find the sun's declination, and note the degree on the brass meridian; then turn the globe around, and all places that pass under that degree will be those required.

EXAMPLES

1. What places have a vertical sun March 20th ?   *Ans.* All places under the Equator.
2. To what places is the sun vertical December 22d ?   *Ans.* To all places under the Tropic of Capricorn.
3. To what places is the sun vertical May 1st ?   *Ans.* To all in latitude 16° N.

PROBLEM V.—To find the meridian altitude of the sun for any day of the year, at any place : Make the elevation of the north or south pole above the wooden horizon equal to the latitude, so that the wooden horizon may represent the horizon of that place; bring the sun's place in the ecliptic to the brass meridian, and the number of degrees on the meridian from the horizon to the sun's place will be the meridian altitude.

EXAMPLES.

1. Find the sun's meridian altitude at New York, June 21st.   *Ans.* 73°.
2. At London, Jan. 27th.   *Ans.* 20°.
3. Rio de Janeiro, September 23d.   *Ans.* 67°.

PROBLEM VI.—*To find the amplitude of the sun at any place, and for any day in the year :* Proceed as in Problem V.; then bring the sun's place to the eastern or western edge of the horizon, and the number of degrees on the horizon from the east or west point will be the amplitude.

EXAMPLES.

1. Find the sun's amplitude at London, June 21st.   *Ans.* 39¾° N.
2. At Quito, September 23d.   *Ans.* 0°.
3. At Philadelphia, July 16th.   *Ans.* 28° N.

PROBLEM VII.—*To find the sun's altitude and azimuth at any place, for any day in the year, and any hour of the day :* Proceed as in Problem V.; then set the index to twelve, and turn the globe eastward or westward, according as the time is before or after noon, until the index points to the given hour. Then, for a vertical, screw the quadrant of altitude over the zenith, and bring its graduated edge to the sun's place in the ecliptic; the number of degrees on the quadrant from the sun's place to the horizon will be the altitude, and the number of degrees on the horizon, from the meridian to the edge of the quadrant, will be the azimuth.

EXAMPLES.

1. Find the sun's altitude and azimuth at New York, May 10th, 9 o'clock A. M. *Ans.* Altitude, 45½° ; azimuth, 72½° E.

2. At London, May 1st, 10 o'clock A. M. *Ans.* Altitude, 47°; azimuth, 44° E.

3. At London, March 20th, 3½ o'clock P. M. *Ans.* Altitude 22°; azimuth, 59° W.

SECTION IV.

DAY AND NIGHT.

108. The *succession of day and night* is caused by the
rotation of the earth on its axis.

As the earth turns on its axis, every place is brought alternately
into the light and into the shade. All places turned toward the sun,
so that its rays can shine upon them, have day; and those turned
away have night, because they are in the earth's shadow.

109. The *comparative length of day and night* at any particular place and time depends upon the sun's declination,
or distance from the equinoctial. When the sun is north of
the equinoctial, all places in the northern hemisphere have
longer day than night, and those in the southern hemisphere,
longer night than day; but when the sun is south of the
equinoctial, this is reversed.

Fig. 46.

In the diagram, let H H represent the
horizon, P P' the axis of the celestial
sphere, E E' the equinoctial; let also S
be the sun in north declination, and S'
in south declination; it will be obvious
that as the earth turns, the sun at S will
appear to move in a diurnal arc, as *a* S *b*,
greater than the nocturnal arc *a c b;* and
at S', the diurnal arc *m* S' *n* will be less
than the nocturnal arc *m o n;* while at E,
in the equinox, the circle of daily motion
described by the sun will be divided
equally by the horizon.

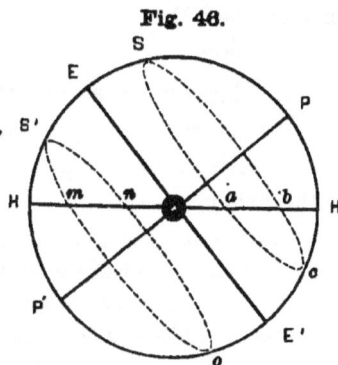

LONGEST DAY AND NIGHT.

QUESTIONS.—108. What causes the succession of day and night? 109. On what does
the length of day and night depend? How does the day compare with the night when
the sun is north of the equinoctial? When south of it? Explain by the diagram.

110. All places under the equator have equal days and nights during the whole year.

a. This will be obvious, if it is remembered that in a right sphere, all the circles of daily motion are perpendicular to the horizon, and equally divided by it; so that whatever the declination of the sun may be, its diurnal arc must always be equal to its nocturnal arc.

b. One half of the year, to places under the equator, the sun is in the north when on the meridian, and the other half in the south; while on the 20th of March and the 23d of September, it is exactly overhead, or *vertical.*

c. Since the sun's declination is never greater than 23½°, no place whose latitude, either north or south, is beyond that limit, can have a vertical sun; and all places within these limits must have a vertical sun twice every year; that is, as the sun moves north and on its return.

111. The small circles parallel to the equator or equinoctial, at the limit of the sun's declination, are called the Tropics; that at the northern solstice is called the TROPIC OF CANCER; that at the southern, the TROPIC OF CAPRICORN.

a. Tropic means turning; and these circles are called tropics, because, when the sun arrives at one, it turns back and goes to the other. They serve to bound that zone of the earth's surface within which the sun can be vertical.

112. Places situated at either of the poles have constant day during the whole six months the sun is in the same hemisphere; and constant night during the six months it is in the other.

a. For in a parallel sphere, the equinoctial coincides with the horizon, and therefore, when the sun is north of the equinoctial, it must be above the horizon, and when south of the equinoctial, below it.

b. Since the altitude of the pole is equal to the latitude, (Art, 82, *a.*)

QUESTIONS.—110. Day and night at places under the equator? *a.* Why? *b.* When is the sun vertical? *c.* What places can have a vertical sun? 111. What are the tropics? How named? *a.* What does *tropic* mean? 112. What places have constant day, and when? *a.* Why? *b.* How often does the sun cross the meridian at these places? Explain by the diagram (Fig. 47).

the distance of the circle of perpetual apparition from the equator is equal to the complement of the latitude; and when the sun is within this circle, there must be constant day : the sun keeping above the horizon, and crossing the meridian twice during the twenty-four hours,—once when it culminates, in the *south* or *north*, according as the place is in *north* or *south* latitude, and once at the opposite part of the circle.

Suppose Z to be the zenith of a place 15° from the north pole, and consequently in 75° N. latitude; H H' being the plane of the horizon, P P the celestial poles, E E the plane of the equinoctial, S the summer solstice, and W the winter solstice : then P H' = 75°, and *a* H' is the circle of perpetual apparition, and H *b* the circle of perpetual occultation. At *d* the sun has 15° of north declination, and at its point of culmination, *a*, crosses the meridian at an altitude of 30° (H E + E *a*); while at the opposite point it just touches the horizon at H';

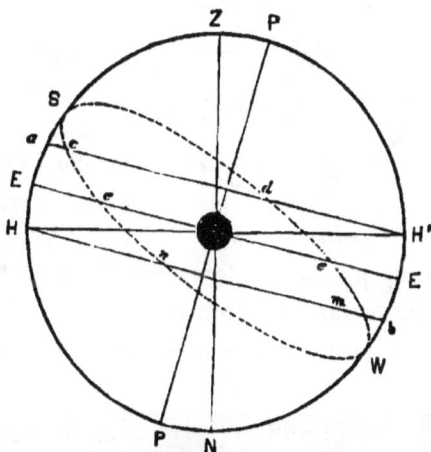

Fig. 47.

CONSTANT DAY AND NIGHT.

H' is obviously the north, because it is toward the pole. Going north from *d* toward the solstice S, and back from *S* to *c*, it is evident that the sun will be within the circle of perpetual apparition, and hence, there will be constant day ; while from *c* to *e*, the equinox, and from *e* to *n*, the circle of perpetual occultation, it will rise and set, its meridian altitude growing less and less until at *n*, it will appear at the horizon for a short time at noon each day, and finally disappear, remaining below the horizon while going south from *n* to W, the winter solstice, and north from W to *m*, where it again crosses the circle of perpetual occultation, and then appears once more above the horizon. There is, therefore, constant day while it is passing over *d S c*, and constant night while passing over *n* W *m*.

c. Since the smallest circle of perpetual apparition or of perpetual occultation reached by the sun extends 66½° from the pole, no place situated within 66½° from the equator, can have constant day, or day of more than 24 hours' duration. Hence, the limit of constant day is the small circle round each pole, 23½° from it. These two parallels are called *polar circles*.

Fig. 48.

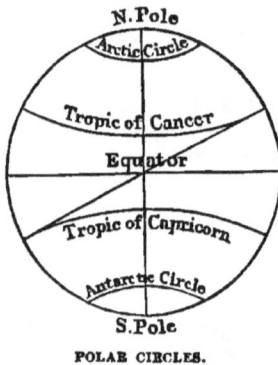

POLAR CIRCLES.

113. The POLAR CIRCLES are two small circles, parallel to the equator or equinoctial, and 23½° from the poles. The one round the north pole is called the ARCTIC CIRCLE; the one round the south pole, the ANTARCTIC CIRCLE.

114. All places within the polar circles have constant day and constant night during a portion of each year; the duration of each being greater or less according as the place is nearer to the pole or farther from it.

The polar circles serve to mark the limit of constant day and night.

Fig. 49.

EARTH AT THE SOLSTICE.

115. When the sun is in either of the *solstices*, all places in the same hemisphere with it have their longest day and shortest night, and those in the other, their shortest day and longest night. When it is in either of the *equinoxes*, the days and nights are of equal length in all parts of the earth, except at, or very near, the poles.

116. The atmosphere of the earth increases the length of the day, both by refracting and by reflecting the sun's rays.

a. When the sun is in either of the equinoxes, one half of it would constantly appear above the horizon of a place at either of the poles, except for refraction, the effect of which is very great at those points; so that the sun, at the equinox, appears wholly above the horizon, thus causing constant day, within one or two degrees of each pole. The general effect of refraction is *to increase the length of the day from six to ten minutes.*

Fig. 50.

EARTH AT THE EQUINOX.

TWILIGHT.

117. When the sun is a short distance below the horizon, its rays fall on the upper portions of the atmosphere, which like a mirror reflect them upon the earth, and thus produce that faint light called *twilight.* The morning twilight is generally called the *dawn.*

Fig. 51.

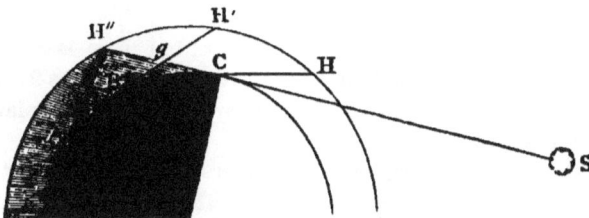

Let A B C represent three places on the earth, and A H″, B H′, C H, their horizons respectively. Suppose S to represent the sun, a little below the horizon, its rays passing through the atmosphere in S C H″; at A, no portion of the visible atmosphere is illuminated, and consequently there is no twilight; at B, the part H″ *g* H is illuminated, and at C, H″ C H; twilight is produced at each of these points.

" When the sun is above the horizon, it illuminates the atmosphere and clouds, and these again disperse and scatter a portion of its light in all directions, so as to send some of its rays to every exposed point from every point of the sky. The generally diffused light, therefore, which we enjoy in the day-time, is a phenomenon originating in the very same causes as twilight. Were it not for the reflective and scat. tering power of the atmosphere, no object would be visible to us out of direct sunshine; every shadow of a passing cloud would be pitchy darkness; the stars would be visible all day, and every apartment into which the sun had not direct admission would be involved in nocturnal obscurity."—*Sir John Herschel.*

118. The duration of twilight varies greatly at different parts of the earth; it is shortest at the equator, and increases toward the poles; near the polar circles and within them, there is constant twilight during a part of each year.

a. At the equator, the duration is 1h. 12m.; at the poles, there are two twilights during the year, each lasting about 50 days. This long twilight diminishes very much the time of total darkness at the poles; for the sun is below the horizon six months, equal to 180 days, and deducting 100 days of twilight, there remain only 80 days, or less than three months, of actual night.

119. Twilight does not cease until the sun is about 18° below the horizon.

a. This is the generally received estimate, but there is considerable uncertainty about it. Some have found it to be as great as 24°; others have reduced it to 16°. There is also a variation in different latitudes; 18° is the mean value.

b. If the earth's atmosphere were more extensive than it is, the twilight would of course be longer, since the sun would not cease to illuminate the higher portions of the atmosphere until more than 18° below the horizon; and if the atmosphere were less extensive, the reverse of this would be the case. Knowing therefore the depression of the sun (18°) requisite for the cessation of twilight, we can calcu late the extent or height of the atmosphere. Thus computed, it is about 40 miles.

QUESTIONS.—118. What is the duration of twilight at different places? Where is there constant twilight? *a.* Duration of twilight at the equator? At the poles? 119. When does twilight cease? *a.* Diversities of estimate? *b.* What can we find by know. ing this fact? *c.* Why is the duration of twilight different at different places? Explain by the diagram (Fig. 52).

c. If the circles of daily motion were at all places equally inclined to the horizon, the duration of twilight would everywhere be the same; since the earth would always have to turn the same amount to bring the sun 18° degrees below the horizon; but the more oblique the circles are, the farther the earth has to turn, and hence the twilight is longer the nearer we go to the poles.

Let the large circle represent the celestial sphere, *e* the earth in the centre; P H the altitude of the pole in one position of the sphere, and P' H its altitude in one less oblique; E E and E' E' the equinoctial in each, and, of course, the direction of the circles of daily motion. In the more oblique sphere, that is, at the place in the more northern latitude, the celestial sphere, or which is the same thing, the earth, would have to turn a distance on the diurnal circle, equal to *e a*, to bring the sun 18° below

Fig. 52.

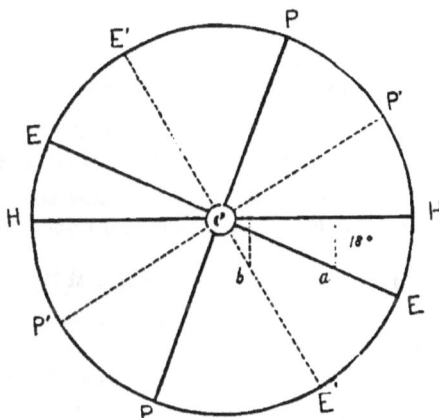

DURATION OF TWILIGHT.

the horizon; while in the other position, the sun would reach the same point of depression when the sphere had turned only *e b.* Thus we see the nearer the perpendicular the diurnal circles are, the shorter the twilight; while the more oblique they are, the longer the twilight.

PROBLEMS FOR THE TERRESTRIAL GLOBE.

PROBLEM I.—To find on what two days of the year the sun is vertical at any place in the Torrid Zone: Turn the globe around, and observe what two points of the ecliptic pass under the degree of the brass meridian corresponding to the latitude of the place; and the days opposite these points in the circle of signs will be those required.

EXAMPLES.

On what two days of the year is the sun vertical at

    1. BOMBAY ?   *Ans.* May 15th and July 29th.

    2. BAHIA ?   *Ans.* Oct. 28th and Feb. 14th.

PROBLEM II.—To find the time of the sun's rising and setting, and the length of the day, at any place, and on any day in the year : Elevate the pole as many degrees as are equal to the latitude of the place, find the sun's place, bring it to the meridian, and set the index to twelve. Then turn the globe till the sun's place is brought to the eastern edge of the horizon, and the index will show the time of the sun's rising ; bring it to the western edge, and the index will show the time of the sun's setting. Double the time of its setting will be the length of the day ; and double the time of its rising, the length of the night.

NOTE.—The globe, of course, only shows this approximatively. A correction would also be required for refraction.

EXAMPLES.

At what time does the sun rise and set, and what is the length of the day and night,

1. At LONDON, July 17th ? *Ans.* Sun rises at 4, and sets at 8 ; length of day, 16 hours ; night, 8 hours.
2. At NEW YORK, May 25th ? *Ans.* Sun rises at 4¾, and sets at 7¼ ; length of day, 14½ hours ; night, 9½ hours.

PROBLEM III.—To find the length of the longest and shortest days and nights at any place not within either of the polar circles : Find, by the preceding problem, the length of the day and night at the time of the northern solstice, if the place be north of the equator, and at the time of the southern solstice, if it be south of the equator ; and this will be the longest day and shortest night. The longest day is equal to the longest night, and the shortest day to the shortest night.

EXAMPLES.

What is the length of the longest and the shortest day

1. At NEW YORK ? *Ans.* Longest day, 14 hours 56 min. ; shortest day, 9 hours 4 min.
2. At BERLIN ? *Ans.* Longest, 16½ hours ; shortest 7½ hours.

PROBLEM IV.—*To find the beginning, end, and duration of constant day at any place within either of the polar circles :* Take a degree of declination on the brass meridian equal to the polar distance of the place, then on turning the globe around, the two points on the ecliptic which pass under that degree will be the places of the sun at the beginning and end of constant day. Find the day of the month corresponding to each, and it will be the times required. The interval between these dates will be the duration of constant day.

Constant night is equal to constant day at a place situated under the corresponding parallel in the other hemisphere. Hence, to find the duration of constant night at a place in north latitude, find the length of constant day at a place having the same number of degrees of south latitude.

EXAMPLES.

Find the beginning, end, and duration of constant day and night at

1. NORTH CAPE. *Ans.* Constant day begins May 14th, ends July 30th ; duration, 77 days. Constant night begins November 25th, ends January 27th ; duration, 73 days.

2. NORTH POLE. *Ans.* Constant day begins March 20th, ends September 23d ; duration, 187 days. Constant night begins September 23d, ends March 20th ; duration, 178 days.

PROBLEM V.—*To find the duration of twilight at any place not within either of the polar circles :* Elevate the pole equal to the latitude, find the sun's place, bring it to the western edge of the horizon, and note the time shown by the index. Then screw the quadrant over the place, and bring its graduated edge to the sun's place ; turn the globe till the sun's place is shown by the quadrant to be 18° below the horizon, and the time passed over by the index will be the duration of twilight.

EXAMPLES.

1. What is the duration of twilight at LONDON, September 23d?
   Ans. 2 hours.
2. What is it at DRESDEN, April 19th? *Ans.* 2 hours 15 minutes.

SECTION V.

THE SEASONS.

120. The SEASONS are the four nearly equal divisions of
the year, which are distinguished from one another by the
comparative length of the day and night, and the difference
in the amount of heat received from the sun.

121. The CAUSES OF THE SEASONS are the inclination of
the axis of the earth to the plane of its orbit, and its revolution around the sun ; and the vicissitudes are regular, that
is, always the same from year to year, because the axis
always points in the same direction, or remains parallel to
itself.

a. These four periods, called *Spring, Summer, Autumn,* and *Winter*
are marked and limited by the arrival of the sun at the vernal equinox, northern solstice, autumnal equinox, and southern solstice, respectively. The following statements will be understood by an inspection
of the accompanying illustration (Fig. 53):

1. **Sun in the Northern Solstice.**—When the sun enters Cancer
(northern solstice), the north pole is presented to the sun ; and summer
is produced in the northern hemisphere, because the rays of the sun
fall *directly* upon that part of the earth ; while winter occurs in the
southern hemisphere, because there the sun's rays are oblique ;

2. **Sun in the Southern Solstice.**—When the sun enters Capricorn
(southern solstice), the south pole is presented to the sun, and summer
occurs in the southern hemisphere, and winter in the northern ;

Fig. 53.

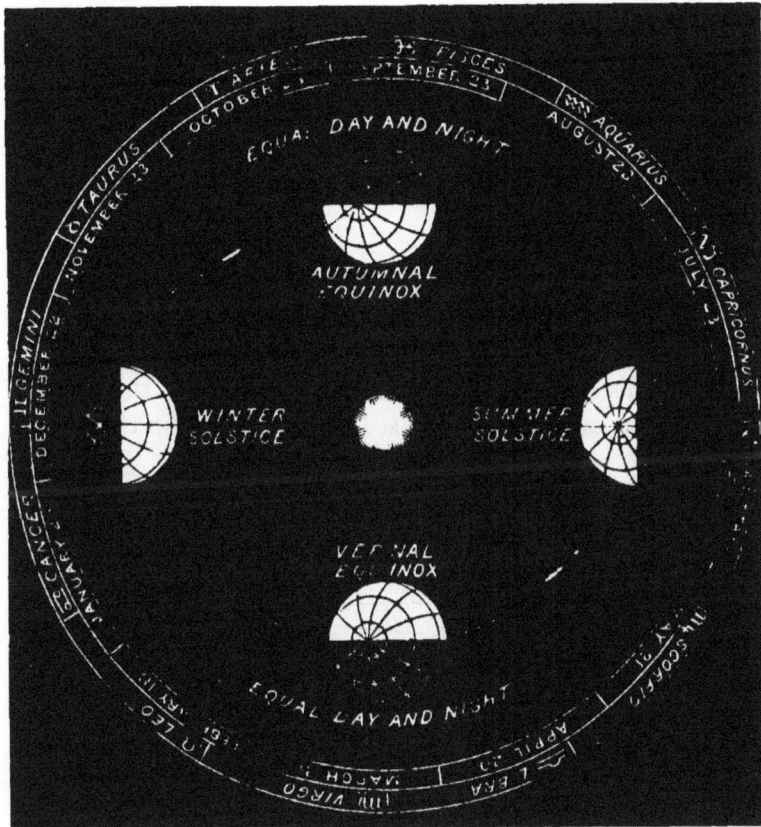

THE SEASONS.

3. **Sun in the Equinoxes.**—When the sun enters either of the equinoxes, the earth's axis leans sidewise to it, and the rays are direct to the equator, and equally oblique on both sides of it. Consequently, there is neither summer nor winter; but spring in that hemisphere which the sun is entering, and autumn in that which it has left;

4. Hence, when the sun enters Aries (vernal equinox), there is spring

QUESTIONS.—When it is at either of the equinoxes? When it enters Aries? Libra?

in the northern hemisphere, and autumn in the southern; and when
it enters Libra (autumnal equinox), the reverse is the case.

b. By the sun's entering a sign, is meant its appearing at the first
point of that sign in the ecliptic; the earth, as seen from the sun,
would appear, of course, at the first point of the opposite sign.

In the inner circle of the diagram (Fig. 53), containing the names of the
months, the dates give the times at which the *earth* enters the corresponding signs in the outer circle. Of course, the *sun*, at these dates enters the
opposite signs.

122. SUMMER is caused by the rays of the sun being more
nearly perpendicular than in the other seasons, so that the
same part of the earth's surface receives a greater quantity
of light and heat. WINTER is caused by the greater obliquity
of the sun's rays, in consequence of which the same quantity of light and heat is diffused over a greater surface.

Fig. 54.

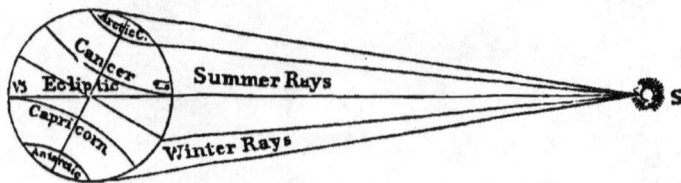

SUMMER AND WINTER RAYS.

In Fig. 54, it will be observed that the same quantity of rays that covers
the north polar circle, when they are direct, covers the whole space from
the antarctic circle to the equator, when they are oblique.

123. The seasons are not precisely of equal length, because the earth revolves in an elliptical orbit, and consequently passes through one half of it in less time than the
other.

a. The perihelion of the orbit is in the 11th degree of Cancer, its
longitude being 100° 21′; so that when the earth is at this point, the

---

QUESTIONS.—*b.* What is meant by the sun's entering a sign? 122. How is summer
caused? Winter? Explain by the diagram. 123. Are the seasons of equal length?
a. Explain the cause.

sun is in the 11th degree of Capricorn, January 1st. Thus, the earth passes its perihelion, and is, consequently, nearest to the sun, in winter; and the time occupied by the sun in going from Libra to Aries, that is, from the beginning of autumn to the beginning of spring, is shorter by about eight days than the time from Aries to Libra, or from spring to autumn again. The seasons are, of course, reversed in the southern hemisphere.

b. The duration of the seasons, respectively, is as follows: Spring, 92.9 days; Summer, 93.6 days; Autumn, 89.7 days; Winter, 89 days. Thus, Spring and Summer contain 186½ days; and Autumn and Winter, 178¾ days; Winter being the shortest season, and Summer the longest.

Fig. 55.

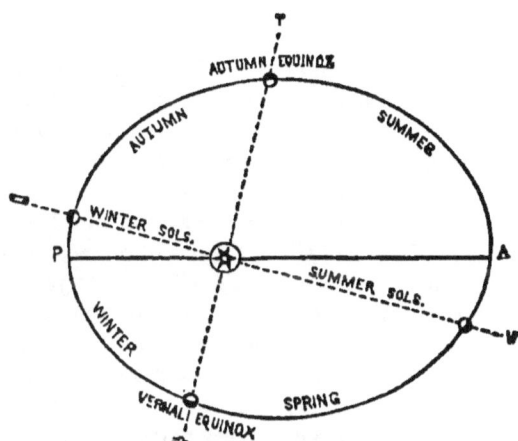

UNEQUAL LENGTH OF SEASONS.

Fig. 55 will render this clear to the understanding of the student. The diagram shows the position of the earth when the sun is at the solstices and equinoxes, respectively, and the unequal portions into which the orbit is divided by the lines joining these points, corresponding to the unequal periods of time mentioned above.

c. **Motion of the Line of Apsides.**—The line of apsides of the earth's orbit does not always remain in the same position in space, but slowly moves toward the east, about 11¾″ every year; hence, making a complete circuit in about 110,000 years. But the equinoxes are moving the other way about 50″ every year (Art. 105), so that the angular distance between the perihelion and the equinox increases annually about 1′ (more exactly, 62″); that is to say, the longitude of the perihelion is about 1′ greater at every successive year.

d. **Length of the Seasons Variable.**—The comparative length of the seasons is, therefore, not the same at different periods. About 6,000 years ago, the aphelion must have coincided with the vernal equinox; and hence, the seasons of summer and autumn must have been equal, and also those of spring and winter; and the former must have been shorter than the latter. About 10,500 years ago, the earth was nearest to the sun in summer, and farthest from it in winter, and the seasons of spring and summer were the shortest, and those of autumn and winter the longest. This would make the summers of the northern hemisphere, according to the calculations of Sir John Herschel, 23° hotter than they now are. [Let the student modify the diagram (Fig. 55) so as to show each of these positions of the line of apsides].

e. **Eccentricity Variable.**—The seasons are also affected, during very long periods, by the variation in the eccentricity of the earth's orbit. At present this is diminishing at the rate of about $\frac{1}{68000}$ of the mean distance in a century; that is, about $36\frac{1}{2}$ miles every year; and as the major axis of the orbit, and, of course, the mean distance, always remain the same, we are, therefore, every year $36\frac{1}{2}$ miles farther from the sun in perihelion, and $36\frac{1}{2}$ miles nearer to it in aphelion, than during the preceding one. If this change continued for ages, the orbit would finally become a circle, and the seasons would be greatly changed; but Lagrange, a famous French mathematician, demonstrated that it takes place only within very narrow limits, at the rate above mentioned. If the eccentricity has continued to diminish for 80,000 years at this rate, at the commencement of that period, it must have been three times as great as at present, or about $4\frac{1}{2}$ millions of miles instead of one million and a half. The aphelion distance must then have been 96 millions, and the perihelion distance 87 millions. Now, the intensity of the solar heat varies inversely as the square of the distance; and the heat of the interplanetary spaces has been estimated at 490° below zero. Hence, if we estimate the average winter heat at 39°, the amount of heat received from the sun must be 529°; and $96^2 : 93^2 :: 529° : 496°$. Hence, if the aphelion distance were 96 millions of miles instead of 93 millions, the average winter heat would be reduced to 6°, or 26° below the freezing point.

124. The DIFFERENCE OF TEMPERATURE in the seasons is

---

not only dependent upon the direction of the sun's rays, but also upon the comparative duration of day and night. Thus, summer occurs when the days are longest, and winter when they are shortest.

a. All parts of the earth's surface are not affected alike by the circumstances which produce the seasons. Those parts of the earth at which the sun may be vertical have the greatest heat; those parts at which there may be constant night have the greatest cold; and the parts between these have a degree of heat and cold not so extreme as either. Hence, the earth's surface has been divided into five portions, called *Zones.*

b. The boundaries of the zones must be the circles which limit the declination of the sun, north and south, and those within which there may be constant day or night; that is, the *tropics* and *polar circles.*

125. The ZONES are the five divisions of the earth's surface bounded by the tropics and polar circles. They are called the Torrid, North Temperate, South Temperate, North Frigid, and South Frigid Zones.

126. The TORRID ZONE includes the space between the tropics, the equator passing through the middle of it. It is 47 degrees wide.

127. The TEMPERATE ZONES are those which are included between the tropics and polar circles. The northern is called the North Temperate Zone; and the southern, the South Temperate Zone. Each is 43 degrees wide.

Fig. 56.

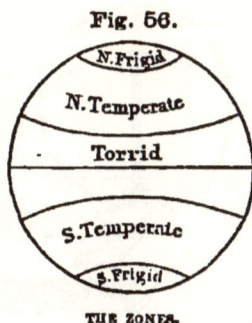

THE ZONES.

128. The FRIGID ZONES are those included within the polar circles. That in the arctic circle is called the North Frigid Zone;

that in the antarctic circle, the South Frigid Zone. Each extends 23½ degrees from the pole, and is 47 degrees across.

SECTION VI.

THE FIGURE AND SIZE OF THE EARTH.

129. THE FIGURE OF THE EARTH is that of an oblate spheroid, differing but slightly from a perfect sphere.

a. **Proofs that the Earth is Spheroidal.**—Several and diverse proofs may be given to establish this fact.

1. *The effect of the centrifugal force would necessarily give it this form;* for, since this force causes bodies to fly off from the centre of motion, the water, or any other yielding materials of which the earth is composed, would recede as far as possible from the axis of rotation, and thus passing from the poles to the equator, cause the earth to bulge out at those parts. Sir Isaac Newton, from this consideration, very nearly ascertained the amount of oblateness in the earth's figure, before any actual discovery of it had been made.

Fig. 57.

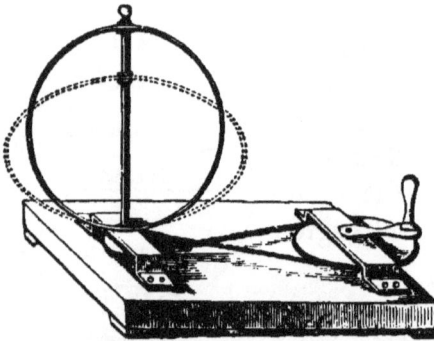

This change in the form of a rotating body may be illustrated by an apparatus represented in Fig. 57. This consists of one or more circular hoops of an elastic material, fastened at the lower end of the axis, but free to move up and down, at the upper end. When set in rapid rotation, they lose their circular form and are bulged out at the points farthest from the axis, so as to become elliptical in form.

2. *The attraction exerted by the earth at its surface is less at the equator than at any other part, and increases as we go from the equator toward either of the poles.* This is shown by a pendulum's vibrating

less rapidly at the equator than at places nearer the poles; and this can be accounted for only by supposing that the equatorial parts of the earth are the farthest from its centre, and the poles the nearest to it; since the attraction of gravitation diminishes as the distance increases.

3. *The length of a degree on the meridian is different in different latitudes,* showing a variation in the curvature of the earth's surface at different parts. If the earth were an exact sphere, the meridians would be perfect circles, and consequently of the same curvature at every part; hence, if we find, by exact measurement, that the curvature is not the same, we know that they are not exact circles. This is what has been ascertained. The length of a degree on the meridian has been measured at different latitudes; and it has been found that it is longer the nearer we go to the poles, showing that the earth is flattened at these parts.

Let the ellipse, Fig. 58, represent the form of the earth. Since the curvature at P is much less than that at E, the radius of the curve *a b* will be longer than that of *c d;* hence, if the angle *a o b* is equal to the angle *c m d,* the arc *a b* which is farther from the centre than *c d,* must be the longer. Of course, this would be equally true of an angle of 1°; and thus, the arc subtending one degree of angular measurement at the poles must be longer than the corresponding arc at the equator, if the earth is spheroidal.

Fig. 58.

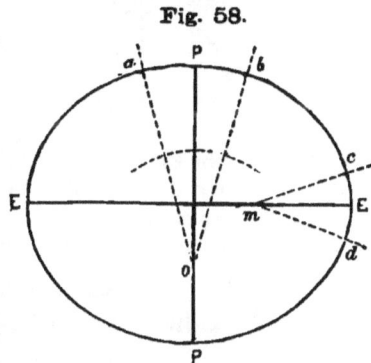

b. **To Find the Size of the Earth.**—The angular distance of two places situated under the same meridian, measured from the earth's centre, is the arc of the meridian contained between the places. This angle is found by observing the change of position, with respect to the horizon or zenith, which a star appears to undergo when viewed from two different points on the earth's surface, one being exactly north of the other. The apparent displacement of the star is the angular distance, or meridian arc, contained between the two places. Then, having measured the distance in *miles* between the places, we can find by a simple proportion, the circumference of the earth. For, suppose the angular distance is found to be $2\frac{1}{2}°$, and the actual distance 172.76 miles; then $2\frac{1}{2}° : 360° :: 172.76$ miles : 24,877 miles. This must be the circumference of the earth; and dividing 24,877 miles by 3.1416, the ratio of the circumference to the diameter, we obtain its diameter.

Fig. 59.

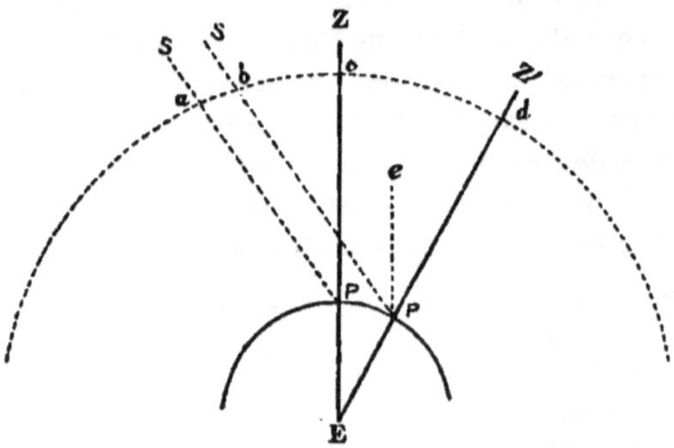

To understand why a change in the place of the spectator causes a displacement of the star, let E (Fig. 59) represent the centre of the earth, P and P places on the earth, Z and Z′ the zenith of each respectively, S, the *direction* of a star situated at an immense distance beyond. At P, the zenith distance of the star is *a c*, or the angle S P Z; at P, the other place, it is *b d*, or the angle S P Z′, greater than S P Z by the angle *e* P *d*, which is equal to the angle P E P. Thus, the star appears farther from the zenith Z′ than from Z at P by the arc of the meridian, P P.

130. The oblateness of the earth's figure is equal only to $\frac{1}{300}$ part of its diameter, or $26\frac{1}{2}$ miles.

a. So small is this variation from an exact sphere, that if a body were made of the precise form of the earth, having its longest diameter three feet in length, the shortest would be only one-eighth of an inch less,—an amount entirely imperceptible.

b. The longest diameter of the earth is $7{,}925\frac{1}{2}$ miles; the shortest diameter 7,899; the mean diameter 7,912 miles.

131. The spheroidal figure of the earth is the cause of the precession of the equinoxes.

a. **Precession Explained.**—For since this excess of matter at the equator is situated out of the plane of the ecliptic, the attraction of

QUESTIONS.—130. What is the degree of oblateness of the earth? Illustration? *b.* What are the exact dimensions of the earth? 131. What does the spheroidal figure of the earth cause? *a.* Explain how precession is caused?

the sun and moon acts obliquely upon it, and thus tends to draw the planes of the equinoctial and ecliptic together; which tendency, by the rotation of the earth on its axis, is converted into a sliding movement, as it were, of one circle upon the other, both preserving very nearly the same inclination.

Fig. 60.

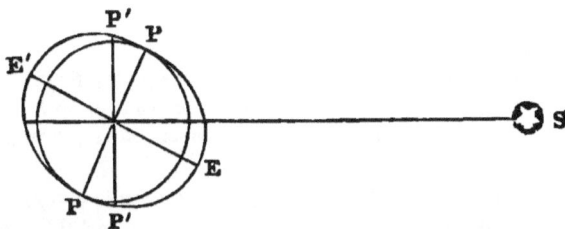

Thus (Fig. 60) the attraction of the sun, acting obliquely upon the protuberance, or excess of matter, at E and E', tends to draw it toward the plane of the ecliptic; and this it would finally accomplish were the earth's rotation suspended; so that the plane of the equator would be made to coincide with that of the ecliptic. But the effect is a sliding of the equator over the line of the ecliptic, and thus a change of the points of intersection.

b. **Revolution of the Poles.**—Since the equator moves round on the ecliptic, the poles of the earth must revolve around those of the ecliptic, and consequently change their apparent position among the stars. Hence, the star which is now so near the north celestial pole will not always be the pole-star; but in about 13,000 years, that is, one-half the period of an entire revolution, will be 47° from it.

c. **Why the Equinoctial Points move toward the West.**—It may not be obvious why the equinoctial points move toward the west; but perhaps the following diagram and explanation will render it clear:

Let E E (Fig. 61) represent the equator, and *e e* the ecliptic, A the first degree of Aries, or vernal equinox; *a b* the amount of force exerted to draw the equator toward the ecliptic in a given time, and *a d* the amount of rotation performed in that time. By the principle of resultant motion, the excess of matter and, of course, the earth with it, would move in the

---

Fig. 61.

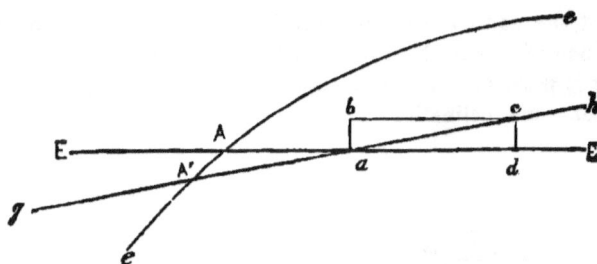

diagonal *a c*, thus changing the direction of the equator from E E to *g h*, and causing the point of intersection to *recede* from A to A'. It will be obvious that the angle of inclination at A must be very nearly equal to that at A'.

d. **Obliquity of the Ecliptic Variable.**—There is a very slow diminution of the obliquity of the ecliptic, amounting to 46¼" in a century. At present (1867), the obliquity is 23° 27' 24". The limit of the variation is 1° 21', to pass through which arc it requires about 10,000 years.

SECTION VII.

TIME.

132. The apparent motions of the sun and stars, caused by the real motions of the earth, afford standards for the measurement of time.

133. The time which elapses between a star's leaving the meridian of a place until it returns to it again is called a SIDEREAL* DAY.

a. This is the time of one complete revolution of the celestial sphere, and is the exact period of one rotation of the earth on its axis. It is an absolutely uniform standard, having undergone not the slightest appreciable change from the date of the earliest recorded

---

* From the Latin word *sidus*, which means a *star*.

---

observations. Indeed, it is the only absolutely uniform motion observed in the heavens.

134. A SOLAR DAY is the period which elapses from the sun's leaving the meridian of a place until it returns to it again.

a. As the sun is constantly changing its place among the stars, owing to the annual revolution of the earth, this period must be longer than a sidereal day ; for the sun having moved toward the east during the time of a rotation, the earth must turn farther in order to bring the place again into the same relative position with the sun. This will be understood by examining the annexed diagram.

Let 1 represent the earth in one position of its orbit, and 2 the position to which it advances during one day ; P, the place at which the sun is on the meridian at 1; P', the same place after one complete rotation, as shown by the parallel P' S. It will be evident that in order to bring P' under the meridian, so that the sun may appear to cross it, the earth will have to turn a space represented

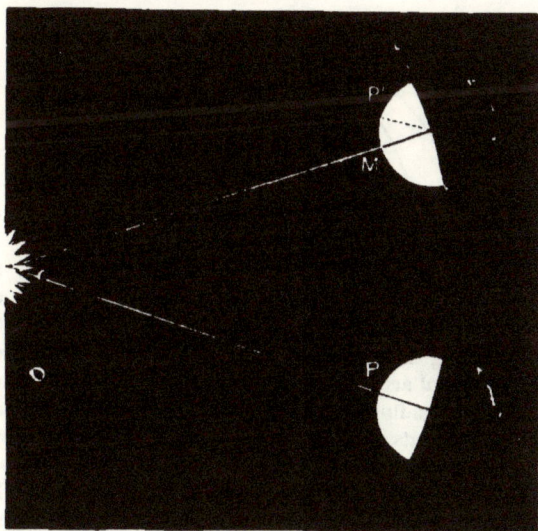

Fig 62.

by the arc P' M, which will make the solar day so much longer than the sidereal day.

135. The solar day exceeds the sidereal day by an average difference of four minutes.

136. Owing to the variable motion of the earth in its orbit, and the obliquity of the ecliptic, this difference is not the same throughout the year; and consequently the solar days are of unequal length.

Why the Solar Days are Unequal.—The first cause assigned for the inequality of the solar days will be easily understood, by referring to Fig. 62; since it will be at once apparent that the length of the arc P′ M must depend upon the length of the interval between 1 and 2. If these intervals vary, the arcs which represent the excess over a rotation turned by the earth in order to bring the sun on the meridian, must also vary, and in the same proportion. Hence, they must be longest when the earth is in perihelion, and shortest when it is in aphelion.

Fig. 63.

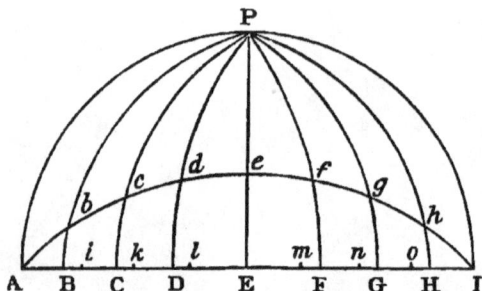

The second cause, namely, the obliquity of the ecliptic, needs an independent illustration:—Let A P I (Fig. 63) represent the northern hemisphere; A E I the equinoctial, and A e I the ecliptic. Let the ecliptic be divided into equal portions, A b, b c, c d, etc., and draw meridians through the points of division, intersecting the equinoctial in B, C, D, etc. The divisions of the ecliptic will be equal arcs of longitude, and the divisions of the equinoctial will be the corresponding arcs of right ascension, and hence passed over by the sun in equal periods of time. These arcs of right ascension, it will be apparent, are not equal; for A b, which is oblique to A B, must subtend a smaller arc, A B, than d e which is nearly parallel to its arc D E. Thus the arcs of right ascension are shortest at the equinoxes, and longest at the solstices; while the divisions coincide at all these four points.

137. A MEAN SOLAR DAY is the average of all the solar days throughout the year. It is divided into twenty-four hours, and commences when the sun is on the lower meridian, that is, at midnight.

a. Because used for the general purposes of civil and social life, it is also called the *civil day.* Clocks are regulated to show its beginning and end, and the equal division of it into hours, minutes, and seconds. As already stated, it is four minutes longer than a sidereal day.

b. If the solar days were equal in length, the sun would always be on the meridian at 12 o'clock ; that is, *apparent noon* would coincide with *mean noon*—the noon of the clock. But this is not the case, and therefore to make the observed noon, as indicated by the sun, correspond with the noon of the clock, a correction has generally to be made, either by adding or subtracting a certain amount of time. This correction is called the *equation of time.*

138. The EQUATION OF TIME is the difference between apparent and mean time ; that is, the difference between time as shown by the sun, and that shown by a well-regulated clock.

a. The unequal motion of the earth in its orbit causes the sun to be in advance of the clock from aphelion to perihelion, that is, from July 1st to January 1st; and behind it from January 1st to Jul·· 1st ; while they both coincide at those points. The obliquity of the ecliptic causes the sun to be in advance of the clock from Aries to Cancer, behind it from Cancer to Libra, in advance again from Libra to Capricorn, and behind again from Capricorn to Aries ; and makes them both agree at those four points. To verify this let the student examine Fig. 63. When these two causes act together, as is the case in the first three months and the last three months of the year, the equation of time is the greatest.

139. The equation of time is greatest in the beginning of November, the sun being then about 16¼ minutes in advance of the clock.

a. Hence, to deduce true noon from apparent noon, at that time it is necessary to subtract 16¼ minutes from the observed time. The sun is at the greatest distance behind the clock about February 10th, the equation being then 14½ minutes, and, of course, to be added, in order to find the correct time.

140. Mean and apparent time coincide four times a year, namely; April 15th, June 15th, September 1st, and December 24th. The equation of time then becomes nothing.

b. To Find the Equation of Time by the Globe.—The part of the equation of time that depends upon the obliquity of the ecliptic can be found by the globe, in the following manner :—Bring the sun's place in the ecliptic to the brass meridian, and find its longitude and right ascension ; the difference reduced to time (counting four minutes to a degree), will be the equation. If the right ascension exceed the longitude, the sun is slower than the clock ; if the longitude exceed the right ascension, the sun is faster than the clock.

Thus, on the 28th of January, the longitude of the sun is about 308°, the right ascension 310¼° ; hence the sun is 10 minutes slower than the clock.

QUESTIONS.—What is the equation of time October 19th ? *Ans.* Sun 10 minutes faster than the clock.
What is it August 13th ? *Ans.* Sun 8 minutes slower than the clock.

141. A SIDEREAL YEAR is the period of time that elapses from the sun's leaving any star until it returns to the same again.

a. This is the true period of the annual revolution of the earth, and is equal to 365 days, 6 hours, 9 minutes, 9 seconds. Owing, however, to the precession of the equinoxes, the sun advances through all the signs, from either equinox to the same again, in a shorter period.

142. A TROPICAL YEAR is the period that elapses from the sun's leaving the vernal equinox until it arrives at it again. It is 20 min. 20 sec. shorter than the sidereal year.

a. Its length is, therefore, 365^d 5^h 48^m 49^s which is the civil year, or the year of the calendar, deducting the 5^h 48^m 49^s ; and as this is very nearly one-fourth of a day, one day is added every fourth year, which makes what is called leap year, or bissextile. The tropical year is sometimes called an *equinoctial* or *solar year.*

b. The sidereal year is not exactly the period which the earth requires to pass from perihelion to perihelion again, since the perihelion is moving slowly toward the east (Art. 123, *c*). This period is called the *anomalistic year.* It is about 4½ minutes longer than the sidereal year.

QUESTIONS FOR EXERCISE.

These questions are to be answered by applying the principles explained in the preceding sections, and without the use of the globe.

1. What is the latitude of the north pole ?

2. What is the latitude of a place under the equator ?

3. New York is about 49½ degrees from the north pole ; what is its latitude ?

4. How many degrees is it from the south pole ?

5. What is the latitude of a place under the Tropic of Cancer ?

6. What under the Antarctic Circle ?   Under the Tropic of Capricorn ?

7. What is the greatest altitude of a heavenly body ?

8. Where is the altitude greatest ?   Where is it least ?

9. If the zenith distance of a body is 15°, what is its altitude ?

10. How many degrees wide is the circle of perpetual apparition in the latitude of New York ?

11. How wide is it at the north pole ?   At the equator ?

12. If the declination of a star is 60° N., does it ever set in New York ?

13. Does it rise in latitude 30° S. ?

14. At what points is the declination of the sun greatest ?

15. At what points is its declination nothing ?

16. What is the right ascension of the sun in the first degree of Cancer ?   What in the first degree of Capricorn ?   In the first degree of Libra ?   In the vernal equinox ?

17. What is the longitude of the sun in the summer solstice ?   In the winter solstice ?   In the autumnal equinox ?

18. When the sun is in either of the equinoxes, what is its meridian altitude in New York ?   In London ?   At Cape Horn ?   At North Cape ?

QUESTIONS.—*b.* What is an anomalistic year ?   Why longer than a sidereal year ?

19. What is the greatest meridian altitude of the sun in New York? What is the least?

20. If the declination of a star is 30° N., what is its meridian altitude in New York? Its zenith distance?

21. What must its declination be to be seen in the zenith at New York?

22. When is it longest day in New York? At Cape Horn?

23. If a star were seen on the meridian 40° from the zenith, what would be its altitude, azimuth, and amplitude?

24. If the meridian altitude of a star in Havana is 50°, what is its declination?

25. What are the amplitude, azimuth, zenith distance, and altitude of a star just rising 15° from the east?

26. What is the right ascension of the sun when its declination is 23½° S.?

27. What is its declination when its longitude is 90°?

28. What is its right ascension when its longitude is 180°?

29. Where is a planet situated when its latitude is 0°?

30. In what position is Mars when it has the same longitude as the sun?

31. At what point of a planet's orbit is the centripetal force greatest? The centrifugal force?

32. If the inclination of the earth's axis had been 30°, how wide would each of the zones have been?

33. If it had been 45°, how wide would the torrid zone have been? The temperate zones?

34. If the earth's axis were perpendicular, where would perpetual summer prevail? Perpetual winter?

35. What would be the seasons, if the earth's axis coincided with the plane of the ecliptic?

36. Is constant day as long at the south as at the north pole?

CHAPTER VIII.

THE SUN.

143. The Sun is the source of light and heat to all the other bodies of the solar system, and the support of life and vegetation on the surface of the earth, or any of the other planets.

All the forces displayed on our planet, whether mechanical, chemical, or vital, spring from the sun and his exhaustless rays; and yet, it is calculated, that the earth, with its limited grasp, only receives the two hundred and thirty millionth part of the whole force radiated and dispensed by this vast and splendid luminary.

144. The greatest distance of the sun from the earth is very nearly 93 millions of miles; and its least distance about 90 millions; making the mean distance, as previously stated, about $91\frac{1}{2}$ millions.

a. **History of its Discovery.**—The distance of the sun from the earth has been, from the earliest times, a subject of close and earnest investigation to astronomers. Ptolemy and those contemporary with him, and in more modern times Copernicus and Tycho Brahe, supposed it to be equal to only 1200 times the radius of the earth, or less than five millions of miles; Kepler thought it to be about fourteen millions of miles; Halley, sixty-six millions; and it was not until the middle of the last century (1769), that any reliable determination of this important fact was reached. This was accomplished by finding the horizontal parallax of the sun by means of observations made at different parts of the earth, of the *transit of Venus*, which took place in that year.

Questions.—143. What is the sun? 144. What is its distance from the earth? *a.* Opinions of various astronomers?

b. When an inferior planet happens to be at or near one of its nodes, at the time of inferior conjunction, it appears like a round black spot on the disc of the sun, and moves across it from east to west. This passage across the disc is called a *transit.* The transits of Venus have been of very great interest because employed to determine the solar parallax. The method will be explained hereafter.

145. The distance of the sun from the earth is ascertained by finding its horizontal parallax. According to a recent determination, this is a little less than 9″.

a. This has been found by a series of observations on Mars, made at the time of its opposition in 1860 and 1862, it being in those years at about its nearest point to the earth. More exactly stated, the solar parallax is 8.94″.

b. It has been already shown (Art. 87, *a.*), that the angle of parallax varies with the distance. The method of determining the distance from the parallax is as follows :

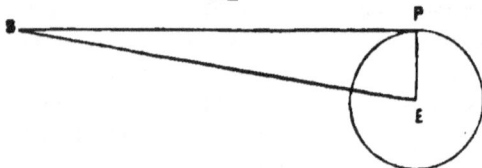

Fig. 64.

Let E (Fig. 64) represent the centre of the earth, P, a place on its surface, and S, the centre of the sun. Then P S E is the angle of horizontal parallax, or the angle which the radius of the earth subtends at the distance of the sun. Now, in every right-angled triangle, such as P S E, the ratio of either side to the hypothenuse depends on the angle opposite the side ; so that however long the sides of the triangle may be, the *ratio* is the same, provided the angle is the same. Hence, as tables have been calculated containing the ratio of every possible angle, we can always find, by referring to these tables, this ratio when we know the angle. In the triangle S P E, S E, the hypothenuse, is the distance of the sun, and P E, the radius of the earth, equal to 3956 miles. The opposite angle P S E, is the horizontal parallax, or 8.94″. For this angle we find the ratio to be about .0000432 ; that is, P E = S E × .0000432 ;

and hence $S E = \dfrac{P E}{.0000432}$; but 3956 ÷ .0000432 = 91,5′4,074, which is about the mean distance of the sun.

c. The ratio of either side of a right-angled triangle to the hypothenuse, dependent upon any particular angle, is called the *sine* of that angle. Thus, the sine of 30° is .5 or ½ ; that is, if one of the angles of a right-angled triangle is 30°, the side *opposite that angle will* be one-half the hypothenuse.

d. Hence, it may be given as a general rule, that *the radius of the earth divided by the sine of the horizontal parallax of any body is equal to its distance from the earth.*

NOTE.—It is important that the student should keep the above definition and rule in memory, as they will be employed in several subsequent calculations.

146. The apparent diameter of the sun, or the angle which it subtends in the celestial sphere, is about 32', or a little more than one-half of a degree.

a. This is the mean value ; the greatest being 32' 36" ; and the least 31' 32". This variation in the apparent size of the sun is caused by the elliptical orbit of the earth ; it being greatest when the earth is in perihelion, and least in aphelion. The apparent diameters of the sun, at different periods of the year, are measures of the different lengths of the radius-vector of the earth's orbit, and thus lead to a knowledge of its exact figure.

b. Since the greatest apparent diameter is 32.6', and the least 31.533', their ratio is as 1.034 to 1 (nearly), and one-half the difference, or .017, is about the eccentricity of the earth's orbit.

147. The actual diameter of the sun is 852,900 miles, or 107¾ times the diameter of the earth.

a. This is found by a calculation based upon the principle of the right-angled triangle, explained in Art. 145. The method is as follows :

Let S (Fig. 65) be the centre of the sun, E the place of the earth. Then S A E is a right-angled triangle, in which the hypothenuse S E is the distance of the sun from the earth, A S the radius of the sun, and the angle A E S

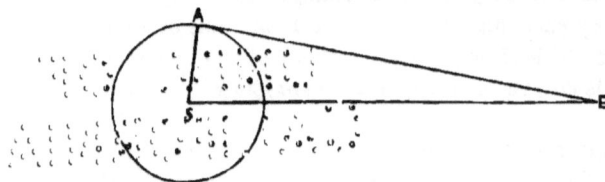

Fig. 65.

one-half the apparent diameter, or 16'. The ratio corresponding to this angle, or the *sine* of the angle, is .00466; hence, 91,500,000 × .00466 = 426,390, the semi-diameter of the sun; and, therefore, the diameter is 852,780 miles, which is very nearly its exact length.

148. The figure of the sun appears to be that of a perfect sphere, no observations having as yet detected any indications of oblateness.

a. **Surface and Volume.**—Since the surfaces of spheres are as the squares of their diameters, and the volumes as the cubes, it follows that the surface of the sun must be 11,620 times that of the earth, and its volume 1,252,000 times; or, in round numbers, one million and a quarter of worlds as large as the earth must be rolled into one to form a body of the bulk of the sun.

149. The mass of the sun is 315,000 times as great as that of the earth.

a. The method of finding this will be explained in a subsequent article. Since the volume of the sun is 1,252,000, while the mass, or quantity of matter is only 315,000, as compared with the earth, it follows that the density of the sun must be only ¼ that of the earth. Now, the earth's density has been found by certain experiments to be about 5½ (5.67) times that of water; hence, that of the sun must be less than 1¼ that of water (1.42).

b. From the comparative lightness of its substance, Herschel infers that an intense heat prevails in its interior, imparting an expansibility sufficient to resist the force of gravitation, which, otherwise, would cause the body to shrink into smaller dimensions.

c. The *volume* of the sun is, as already stated, about 500 times that of all the planets; the *mass* is, however, about 700 times as great.

This shows that the density of the sun is greater than the *average* density of the planets.

150. The sun rotates from west to east on an axis nearly perpendicular to the plane of the ecliptic, the period of rotation being about $25\frac{1}{2}$ days (25^d 7^h 48^m).

151. This is proved by the spots which are seen upon its disc, and which appear to move across it, occupying about two weeks in their passage.

Fig. 66.

A SPOT PASSING ACROSS THE DISC.

a. A particular spot which can be identified by its appearance first appears on the eastern limb, or edge, of the disc, passes across to the western limb, and then disappears; but after about two weeks, re-appears on the eastern limb, completing an entire revolution in about

Fig. 67.

MOVEMENT OF THE SUN AND SPOTS.

$27\frac{1}{2}$ days. But this must be longer than the period of a rotation, because the earth is moving in its orbit in the same direction. When, therefore, the earth has completed one revolution, the number of revolutions of the spots will be one less than the actual number of rotations of the sun for the time. Hence, $365\frac{1}{4}$ days $\div 27\frac{1}{4}$ days $= 13.4$, revolutions of spots; and $13.4 + 1 = 14.4$, rotations of the sun; therefore, $365\frac{1}{4}$ days $\div 14.4 = 25\frac{1}{3}$ days, the time of one rotation. (See Fig. 67).

b. It may appear singular, at the first view, to infer an eastward rotation of the sun from an apparent westward motion of the spots; but it must be remembered that the sides of the sun and earth presented to each other at any time are moving in opposite directions in space, while both bodies move in the same direction in circular motion.

c. **Discovery of the Spots.**—The discovery of spots on the solar disc is noticed in history as early as 807 A. D.; but their true appearance and extent were unknown until the invention of the telescope, in the beginning of the 17th century, at which time (in 1611) they were attentively observed by Galileo and others. In recent years, the sun has received a very great deal of attention from astronomers, and many interesting facts have been made known respecting its appearance and physical constitution.

152. The inclination of the sun's axis to the ecliptic is $7\frac{1}{3}°$; and, in consequence of this inclination, the spots appear to move across the disc in lines of various directions and form, sometimes being straight and sometimes curved.

Fig. 68.

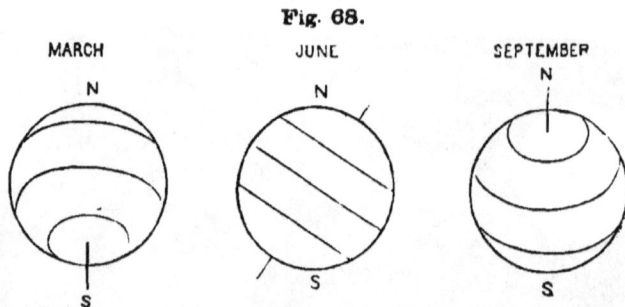

APPARENT PATHS OF SOLAR SPOTS.

Fig. 68 illustrates this. In March, when the south pole is presented to the spectator, the paths assume the appearance indicated in the first circle; in June, they are straight and oblique, because the observer is in the plane of the sun's equator; while in September, the observer being north of its equator, the north pole is turned toward him, and they are as represented in the third circle. [The inclination of the axis is exaggerated in the diagram.]

153. APPEARANCE OF THE SPOTS.—When the spots are examined by means of a telescope, they present the appearance of irregular black patches surrounded with a dusky border or fringe, the whole sometimes encompassed with a bright surface or ridge.   The black portion in the centre is called the *umbra* or *nucleus ;* the dusky border, the *penumbra ;* and the bright surfaces seen around the spots, or by themselves on other parts of the disc, are called *faculæ.*

Fig. 69.

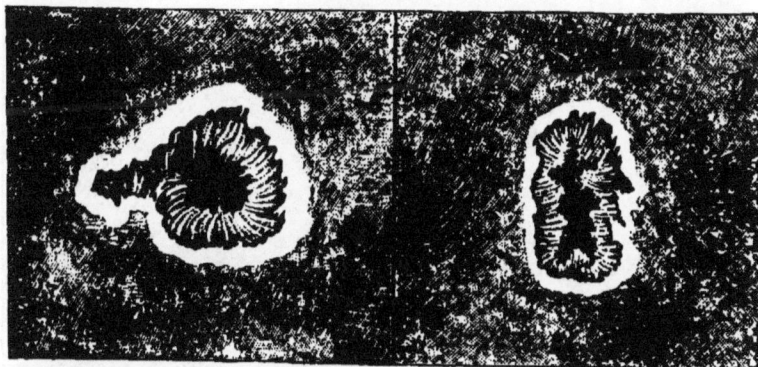

SOLAR SPOTS.

a. Sometimes the nucleus is absent ; and sometimes spots are seen without any penumbra. The nucleus is not of a uniform blackness, but generally contains an intensely black spot in the centre. These spots usually appear in clusters, numbering from two to sixty or seventy, or even many more.

154. VARIABILITY OF THE SPOTS.—The solar spots constantly undergo very great changes in number, form, size, and general appearance.

QUESTIONS.—153. Explain the appearance of the spots. *a.* What diversity in their appearance ? 154. What changes do they undergo?

a. Sometimes the sun's disc will be entirely free from them, and will continue so for weeks and months; at other times, they will burst forth and spread over certain parts of it in great numbers. After twenty-five years of continued observations, M. Schwabe, a German astronomer, discovered that there was a periodical increase and decrease of the number and size of the spots; and Prof. Wolf, of Zurich, by comparing the observations made during the last hundred years, has shown that this period has varied between 8 and 16 years. These periods are thought by some to depend upon physical influences exerted by some of the planets, particularly Venus and Jupiter, when in certain positions of their orbits.

Fig. 70.

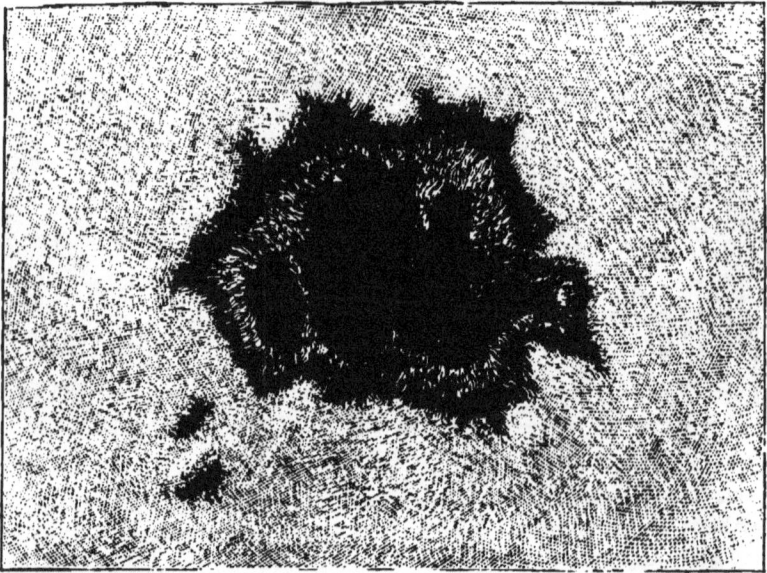

SUN-SPOT, JULY 29, 1860, SHOWING THE "WILLOW-LEAF" STRUCTURE.

b. The spots are mostly confined to two zones parallel to the equator, and extending from 5° to 35° from it; and they appear to have a tendency to arrange themselves in lines parallel to the equator.

c. The duration of single spots is also very variable. A spot has been seen to make its appearance and vanish within twenty-four hours; while others have continued for nine or ten weeks, without much change of appearance.

d. Their magnitude also presents very great diversity. Spots are not unfrequently seen that subtend an angle of more than 60'', or nearly seven times the sun's horizontal parallax; the diameter of such spots must therefore be more than 25,000 miles. A spot in June, 1843, continued visible to the naked eye for a whole week, its length being estimated at 74,000 miles. One observed in 1839, by Capt. Davis, had a linear extent of 186,000 miles.

Fig. 70 represents a large spot as seen and drawn by Mr. Nasmyth, an English astronomer, in 1860. It shows the umbra, penumbra, the latter arching the former as well as surrounding it, and also the dotted or mottled surface of the sun, as seen through a powerful telescope. The penumbra presents the appearances to which Mr. Nasmyth has applied the name of "willow leaves," from their fancied resemblance to such objects.

155. Theories as to the Physical Constitution of the Sun.—The most generally received hypothesis as to the nature of the sun is that it is an opaque body surrounded by an atmosphere of luminous matter, and that the spots are openings in the atmosphere, through which the dark body of the sun becomes visible.

a. This hypothesis was first advanced by Dr. Wilson, of Glasgow, in 1769. In 1793, Sir William Herschel suggested the hypothesis that two atmospheres encompass the sun; the first or lower one being formed of a partially opaque or cloudy stratum reflecting light, but emitting none of itself; and the second consisting of luminous matter, which is the source of the sun's light, and gives to the disc its form and limit. This luminous atmosphere has been sometimes called the *photosphere.*

b. The existence of a *third* atmosphere, very nearly transparent, and extending a great distance above the photosphere, is clearly indicated by the diminished brightness of the sun's disc toward the edges.

c. Wilson's and Herschel's hypotheses, as developed and modified

Questions.—*d.* Their magnitude? 155. What generally received hypothesis as to the cause of the spots? *a.* By whom advanced? *b.* What evidence of a third atmosphere? *c.* How do Wilson's and Herschel's hypotheses explain the phenomena? Cause of the openings?

by more recent observers, explain all the phenomena of the spots. The black umbra is the body of the sun, while the penumbra is the non-luminous atmosphere, or cloudy stratum, rendered visible by the larger opening in the photosphere above it. When this opening is smaller, no penumbra is visible; and when there is no opening in the cloudy stratum, no black nucleus is visible. These openings or rents are supposed by Sir John Herschel to be caused by changes of temperature, in a manner similar to the production of tornadoes and other agitations of the earth's atmosphere.

156. The spots and other appearances on the sun's disc indicate, without doubt, the existence of a luminous atmosphere, consisting of gaseous matter in an incandescent state,—like the flame of an ordinary gas-burner,—and another atmosphere, also gaseous, and almost perfectly transparent, extending to a considerable distance beyond.

a. The gaseous character of the atmosphere, denied by Sir William Herschel, seems to have been conclusively proved by M. Arago, by means of an ingenious application of the principle of polarized light. M. Faye estimates the height or extent of the photosphere at 4,000 miles.

b. **Kirchhoff's Hypothesis.**—A simpler hypothesis than Wilson's and Herschel's has within the last five years been advanced by Kirchhoff, a German physicist, and others, to account for the phenomena of the spots, consistently with the established facts, as above stated. According to this hypothesis the nucleus of the sun is an incandescent, solid or liquid mass, the vapors arising from which form the atmospheres, the denser and lower one being luminous from the incandescent particles that float in it. Changes of temperature in this atmosphere give rise to tornadoes and other violent agitations; and descending currents produce the openings, which are dark because filled with clouds of various degrees of condensation. This theory, and the experiments upon which it is based, are receiving, at present, much attention from astronomers and physicists; and there is reason to believe, that when fully developed, it will entirely supersede the cumbrous and therefore improbable hypothesis so long and so ingeniously sustained.

QUESTIONS.—156. What is certainly indicated by the phenomena? *a.* Gaseous character of the atmosphere? *b.* Explain Kirchhoff's hypothesis.

Fig. 71.

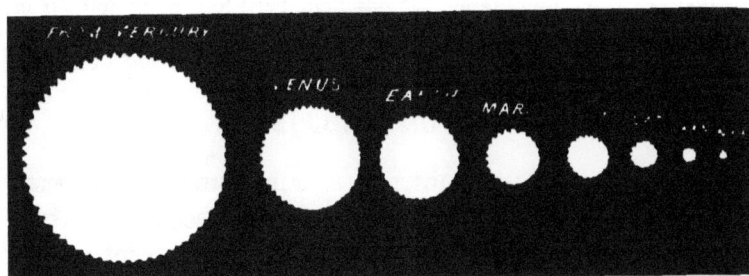

APPARENT MAGNITUDES OF THE SUN.

157. The apparent diameter of the sun at each of the planets diminishes in proportion as the distance increases. Thus, at Mercury, it is $2\frac{1}{2}$ times as great as at the earth; but at Neptune, only $\frac{1}{30}$ as large.

a. The surface of the solar disc at Mercury must therefore be about 6,000 times as great as at Neptune, and the intensity of its light and heat in the same proportion.

b. Various experiments seem to show that the *light of the sun* at the earth is equal to that of 600,000 full moons; (Wollaston estimated it at 800,000.) The light of the sun at Neptune must therefore be equal to about 670 times that of the full moon at the earth. The *electric light* is the only light that approximates in intensity to the light of the sun.

c. The *intensity of heat* at the surface of the sun has been estimated to be 300,000 times that received at any point of the earth's surface. Sir John Herschel supposes that it would be sufficient to melt a cylinder of ice 45 miles in diameter, plunged into the sun, at the rate of 200,000 miles a second.

158. In addition to the rotation on its axis, the sun appears to have a progressive motion in space, revolving with all its attendant bodies around some remote star or centre.

a. The point to which it is tending has, from a vast number of observations made by different astronomers, been located in 260° 20′ of right ascension, and 33° 33′ of declination. Its annual velocity is supposed to be about 160 million of miles.

THE ZODIACAL LIGHT.

159. The ZODIACAL LIGHT is a faint luminous appearance, of the form of a triangle or cone, seen at certain seasons of the year, in the evening at the western, and in the morning at the eastern horizon.

Fig. 72.

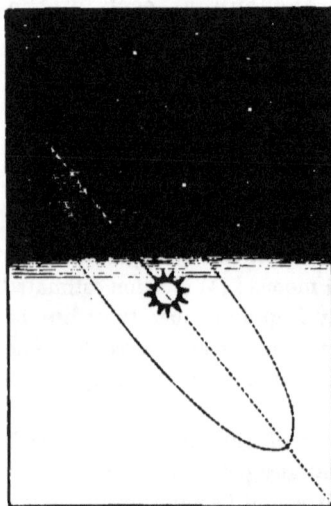

a. Its color is a faint white, tinged with yellow at the base, and fading away toward the apex, which is not sharp, but obtuse, or rounded. It extends obliquely from the horizon, in the plane of the sun's equator, and hence, very nearly in that of the ecliptic; the distance of its apex from the sun varying from 40° to 100° or more. Its breadth at the horizon also varies from 8° to 30°.

b. It is seen most distinctly in March and April after sunset, and in September and October before sunrise; because, at those times, the ecliptic is most nearly perpendicular to the horizon. In tropical regions it is more conspicuous than in the higher latitudes, and has been seen at midnight at both sides of the horizon at once, extending upwards so as almost to form a luminous arch.

It appeared thus to Chaplain Jones of the U. S. Navy, who, from 1853 to 1857, made a long and careful series of observations of it at the equator and between the tropics. He thought the observed phenomena proved it to be a nebulous ring encompassing the earth. Humboldt, in the same latitudes, also saw the double appearance of this light.

c. **Cause of the Zodiacal Light.**—Various hypotheses have been suggested to account for the zodiacal light ; that most generally received at present is, that it is a nebulous mass of great tenuity, of the shape of a lens, encompassing the sun at its equator, and extending sometimes beyond the orbit of the earth.

d. It must therefore sometimes envelop the earth in the plane of the ecliptic ; and consequently, to a person situated at the equator, or at either of the solstices, when the sun is in the other, would necessarily appear, about the time of midnight, at both sides of the horizon ; while, farther north or south, it would disappear at that time, because viewed at a lower altitude, and through its narrowest part ; and would there be visible only in the evening, near the sun, where the line of view would penetrate it at its greatest thickness.

e. Professor Norton regards it as made up of " streams of particles continually flowing away from the sun, under the operation of a force of solar repulsion due to disturbances occasioned by the planets in the magnetic condition of the particles composing the photosphere, and, therefore, arising from the same physical cause as that which produces the spots." He also traces a connection between it and the luminous appearance called the *corona,* seen at the time of a total eclipse of the sun around the obscured disc. The zodiacal light, he thinks, " may vary in brightness from one year to another, with the varying activity of discharge from the sun's surface." By others it has been regarded as a vast ring of meteors circulating about the sun, and finally impinging upon it.

f. **Meteoric Theory of the Sun's Heat.**—This hypothesis of a constant shower of meteoric bodies falling upon the sun, has been used to account for the support of its heat ; for their collision with the sun would necessarily generate an intense heat, just as iron may be heated to any degree by hammering it. It is calculated that bodies of the density of granite falling all over the sun to the depth of 12 feet in a year, and with the velocity which they would acquire (384 miles in a second), would maintain the solar heat. If Mercury were to strike the sun, it would generate an amount of heat equal to all the sun emits in seven years ; while the shock of Jupiter would supply the loss of more than 30,000 years.

CHAPTER IX.

160. The MOON, although one of the smallest bodies in the solar system, is, to us, next to the sun, the most conspicuous, interesting, and important, on account of its close connection with our own planet, and the effects which it produces upon it.

161. The orbit of the moon is elliptical; the point nearest to the earth being called the PERIGEE,* and the point farthest from it, the APOGEE.†

162. Its mean distance from the earth is 238,800 miles; and it is 26,000 miles nearer to us in perigee than in apogee.

a. Its eccentricity is, therefore, 13,000 miles, or about .055 of its mean distance. This is more than three times as great, in proportion, as that of the earth, which is less than .017.

b. To Find its Distance.—The distance of the moon is found by the method and rule explained in Art. 145. The moon's mean horizontal parallax is 57′, the sine of which is .01657: hence, 3956 ÷ .01657 = 238,745; which is very nearly the distance found by exact computation.

c. This is the distance of the moon from the earth's centre; consequently it is about 4,000 miles nearer to a point of the earth's surface

---

* From the Greek words *peri*, meaning *near*, and *gee, the earth*.

† From the Greek words *apo*, meaning *from*, and *gee, the earth*.

---

directly under it ; and, with reference to any particular place on the surface of the earth, its distance varies with its altitude, being greatest at the horizon, and least at the zenith ; that is, about 4,000 miles farther in the horizon than when in the zenith.

Thus (Fig. 73), when the moon is at A, in the horizon, its distance from the place P is A P ; but at B, in the zenith, it is B P ; and A P is obviously greater than B P by nearly the radius of the earth, or about 4,000 miles.

Fig. 73.

d. **Motion of the Apsides.**—The positions of the apogee and perigee in space are determined by noticing when the moon's apparent diameter is greatest and when least. Careful observations of this kind show that these points shift their positions. and that the line of apsides completes a circuit from west to east in ♈ 310¼ᵈ. This is called the *progression of the apsides.*

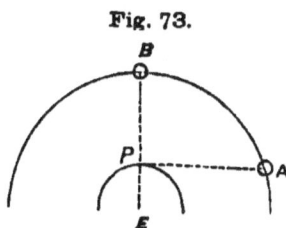

163. The inclination of the moon's orbit to the plane of the ecliptic is about $5\frac{1}{7}°$; consequently, it crosses this plane in two points called the *moon's nodes.*

a. Their positions are ascertained by observing from day to day the distance of the moon's centre from the ecliptic, which is its latitude, and noticing when the latitude becomes nothing. It must then be in one of the nodes ; when it comes from the south, the ascending node, and when from the north, the descending node.

b. **Motion of the Line of Nodes.**—The line of nodes, like the line of apsides, is subject to a change, but in a retrograde direction, or from east to west. It completes a revolution in 18½ years.

164. The mean apparent diameter of the moon is 31¼′, or a little more than half of a degree ; being about the same as that of the sun. The real diameter of the moon is, therefore, 2,162 miles.

a. **The Size of the Moon Calculated.**—For the distance multiplied by the sine of the apparent diameter is equal to the real diameter (Art. 147, *a*). The sine of the apparent diameter is .009055, and $238,800 \times .009055 = 2162.3$, which is the real diameter of the moon.

b. **Surface, Volume, Mass.**—The diameter being very nearly equal to $\frac{3}{11}$ that of the earth, its surface is $(\frac{3}{11})^2$, or $\frac{9}{121}$, or about $\frac{2}{27}$ of the earth's surface; and its volume $(\frac{3}{11})^3$, or about $\frac{1}{49}$ that of the earth. Its *mass* is estimated to be about $\frac{1}{80}$ of the earth's; and consequently its density must be considerably less, about $\frac{3}{5}$.

PHASES OF THE MOON.

165. The moon, when she first becomes visible in the west, is seen as a slender crescent; but from evening to evening her form expands as her angular distance eastward from the sun increases, until when in quadrature, or 90° from the sun, half of her disc is visible. When she has departed so far to the east that she rises just as the sun sets, the whole of her disc is seen, and she is said to be *full.* After this she becomes the waning moon, rising later and later, and growing less and less, until she may be seen in the east as a bright crescent just before sunrise. A short time after this she disappears, and then becomes visible again in the west. These different appearances, called the *phases of the moon,* prove that she revolves around the earth from west to east.

166. When the moon is in conjunction, the dark side being turned toward us, she is called *new moon ;* when she is in quadrature and shows half of her disc, she is called *half-moon ;* when she is in opposition, she is called *full moon.* When she is in quadrature after conjunction, she is said to be in her *first quarter ;* when in quadrature after opposition, in her *last quarter.*

Fig. 74.

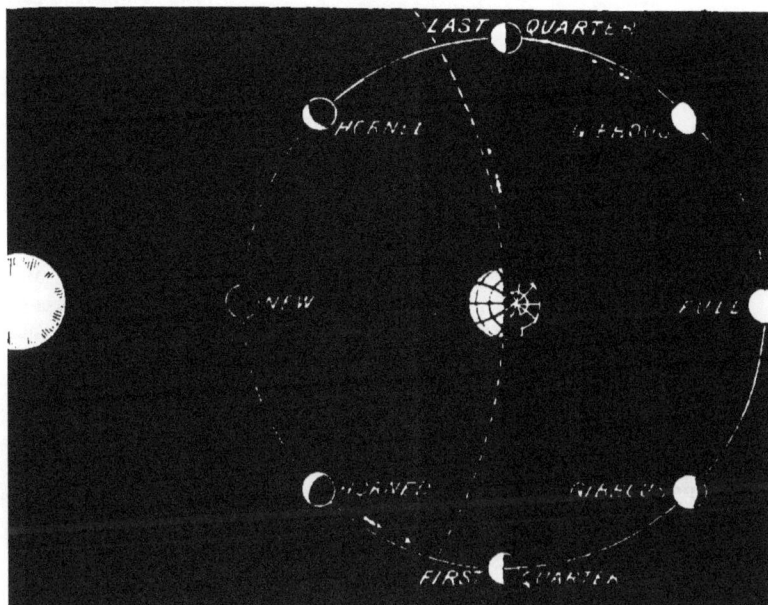

PHASES OF THE MOON

167. When she is between conjunction and quadrature she assumes the crescent form, and is then said to be *horned ;* when she is between opposition and quadrature, she exhibits more than one-half of her disc, but not the whole, and is said to be *gibbous.*

The positions of new and full moon are sometimes called the *syzygies.**

168. The phases of the moon are the different portions of her illuminated surface which she presents to the earth as she revolves around it.

---

* From the Greek word *syzygia*, meaning a *yoking together.*

Fig. 75.

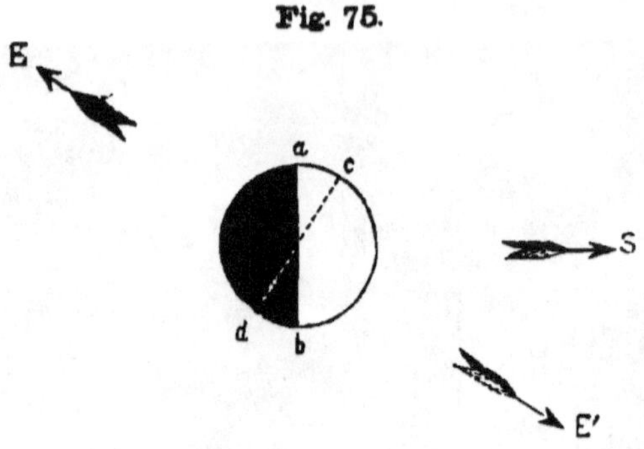

MOON HORNED AND GIBBOUS.

In Fig. 75, let the partially darkened circle represent the moon; S, the direction of the sun; E, the direction of the earth on one side of the moon, and E′, its direction on the opposite side. Then *a b* will represent the line which separates the illuminated and darkened hemispheres of the moon: and *c d*, that which separates the hemisphere turned toward the earth from that turned away from it. At E, *a c* being the only part of the disc visible, the moon appears horned; while at E′, *b c* being visible, the form is gibbous.

a. Hence we can find the time of a revolution of the moon by observing the phases. If the earth were at rest, the time from one new or full moon to the next would be exactly the period of a revolution; but as the earth is constantly advancing in her orbit, when the moon has completed a revolution, she has to move still farther in order to come into the same relative position with the earth and sun.

169. The time from one new moon to the next is $29\frac{1}{2}$ days. This is the synodic period, and is called a *synodical month*, or *lunation*.

a. **Sidereal Period Calculated.**—In a year, or $365\frac{1}{4}$ days, the moon makes $365\frac{1}{4} \div 29\frac{1}{2}$, or $12\frac{45}{118}$ synodic revolutions; but the sidereal, or actual, revolutions of the moon must be *one* more; because each synodic revolution is equal to one sidereal revolution and a part of another, equal, in angular measurement, to the advance of the earth in her orbit during each synodic revolution of the moon. Hence, the moon performs $13\frac{45}{118}$ sidereal revolutions in $365\frac{1}{4}$ days: but $365\frac{1}{4}$ days $\div 13\frac{45}{118} = 27\frac{1}{3}$ days (nearly), which is, therefore, the time of one sidereal revolution.

In Fig. 76, let A B represent the advance of the earth in its orbit, while the moon completes a synodic revolution, that is, moves from C, the position of inferior conjunction, till she arrives at the same relative position

---

QUESTIONS.—*a.* What can we find by the phases? 169. What is a synodical month, or lunation? *a.* How to find the sidereal period? Explain by the diagram.

with the sun at E. But when she reaches this point, she has completed a sidereal revolution, and has also moved from D to E, a distance, it will be

Fig. 76.

SIDEREAL AND SYNODICAL REVOLUTION.

seen, equal in angular measurement to A B; since the arc A B bears the same proportion to the earth's orbit that E D does to that of the moon.

170. Owing to the constant advance of the moon in her orbit, she rises and, of course, arrives at the meridian and sets, about 50 minutes later each successive day.

a. This is the average interval of time between the successive risings of the moon ; for since she moves through the ecliptic in $29\frac{1}{2}$ days, her daily advance is equal to about $12\frac{1}{2}°$; but a place upon the earth's surface moves $15°$ in one hour, and hence, requires nearly 50 minutes to overtake the moon. If the moon's orbit or the ecliptic, since the inclination is very small, always made the same angle with the horizon, this would be the constant interval ; but, in consequence of the obliquity of the ecliptic, this angle continually varies during each lunation.

171. The HARVEST MOON is the full moon that occurs in high latitudes, near the time of the autumnal equinox, in September and October, when she rises but a little later for several successive evenings, and thus affords light for collecting the harvest.

a. By means of the globe, it may be easily shown that the ecliptic is most oblique to the horizon in the signs Pisces and Aries, and least so in Virgo and Libra ; so that when the moon is in the former signs, in this latitude, she rises only about half an hour later, but when in

QUESTIONS.—170. Why does the moon rise later each evening? *a.* Why are the intervals unequal? 171. What is harvest moon? *a.* How to explain this phenomenon?

the latter, more than an hour. This difference is, however, only noticed when the moon happens to be *full* while in Pisces or Aries, and thus rises, for several evenings, in the higher latitudes, but a few minutes later. These full moons must occur, of course, in September and October, when the sun is in the opposite signs, Virgo and Libra. In the former month, the full moon in England is called the *Harvest Moon ;* in the latter, sometimes, the *Hunter's Moon.*

Fig. 77.

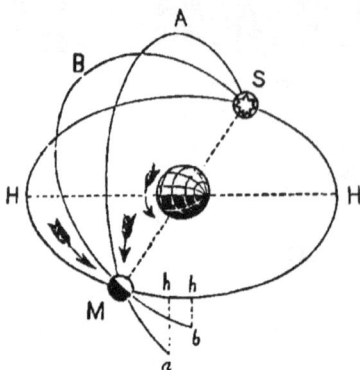

HARVEST MOON.

Let H S H M (Fig. 77) represent the horizon; S, the position of the sun at sunset ; M, the full moon just rising; S A M, the part of the equator, and S B M, the part of the ecliptic above the horizon, the sun being in Libra, the autumnal equinox, and the moon in Aries, the vernal equinox. Since the *southern* half of the ecliptic lies *east* of Libra, it will be evident that in or near this position the ecliptic must make the smallest angle with the horizon ; and consequently, while the moon makes her daily advance in her orbit, M *b*, she only descends below the horizon a distance equal to *h b ;* while, if her orbit made a greater angle with the horizon, as S A M, she would, by advancing through the equal arc M *a*, descend below the horizon a distance equal to *h a*.

b. **In the Polar Regions,** since the full moon must be opposite to the sun, it remains constantly above the horizon ; and during about 15 days passes through its changes without rising or setting, appearing to move around the horizon ; and at the pole, in a circle exactly parallel to it. At the time of the solstice, it is first seen in the west in its first quarter, and continues constantly visible till the last quarter. These brilliant moonlight nights serve partially to compensate the inhabitants of those dreary regions for the long absence of the sun.

c. **Moonlight in Winter.**—The moonlight nights in the temperate latitudes are longer and more brilliant in winter than in summer; especially about the time of the winter solstice. For when the sun is in Capricorn, $23\frac{1}{2}°$ south of the equinoctial, the full moon is in the op-

QUESTIONS.—*b*. The moon as seen at the polar regions? *c*. Moonlight in winter?

posite sign, Cancer, 23½° north of the equinoctial, and therefore culminates at a great altitude ; and, if she happens to be also at the point of her orbit, 5½° north of the ecliptic, at her greatest altitude, which is equal to the complement of the latitude plus 23½° plus 5½°. In New York, this is 49° + 23½° + 5½° = 77° 38'.

172. Observations with the telescope show that the moon always presents very nearly the same hemisphere to the earth. This proves that it rotates on its axis once during each sidereal month, or 27¼ days.

a. The unassisted eye is able easily to perceive that the dusky spots on the disc of the moon constantly keep in the same relative position and present the same appearance ; and this could not occur if she rotated so as to present in succession different hemispheres to the earth. Just as we infer a rotation of the sun from the apparent motion of the solar spots, so we know that the moon rotates during one revolution around the earth, by the observed fact that the lunar spots have no apparent motion ; since, if the moon performed no rotation, the spots on its disc would move across it from west to east, keeping pace with the moon's motion in the ecliptic, and completing one apparent revolution in 29½ days.

Fig. 78.

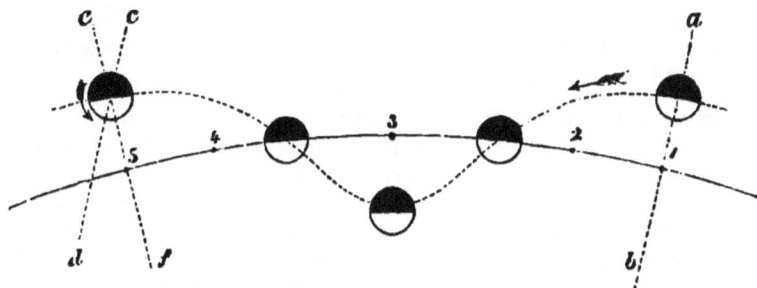

That the moon must perform one rotation during each *sidereal* month, in order to keep the same side turned toward the earth, will be evident from the annexed diagram (Fig. 78). Let the line 1, 2, 3, etc., represent a portion of the earth's orbit, and the dotted curve the real orbit of the moon, as it is carried by the earth around the sun during one lunation. When the earth is at 1, the moon is full; at 2, last quarter; at 3, new; at 4, first quarter; and at 5, full again.   The line *a b* indicates the position of the moon at the commencement of a rotation; and the *parallel line c d*, its position if it had only completed a rotation at the end of the lunation; but it is evident that in order to keep the same face to the earth at 5, it must have turned more than one rotation by the angle contained between *c d* and *e f.*   Hence, during a synodic period, or lunation, the moon performs more than one rotation, which she completes in a sidereal period, or 27¾ days.

173. The real orbit of the moon, as she is carried by the earth around the sun, crosses the earth's orbit every 14½°, but departs so little from it that it is always *concave* to the sun.

a. It will be evident from Fig. 78, that the moon crosses the earth's orbit twice during each lunation, or 29½ days; but there are nearly 12½ lunations in a year; hence the moon must cross the earth's orbit 25 times during one year; and 360' ÷ 25 = 14½° (nearly).

b. **The Lunar Orbit.**—The orbit of the moon, if correctly represented in relation to that of the earth, would present the appearance of a continuous curve, never crossing itself, and so slightly deviating from the earth's orbit as, unless drawn on a very large scale, scarcely to be distinguished from it.   This will be evident when it is considered that the moon's distance from the earth is only about $\frac{1}{400}$ of the earth's distance from the sun.   Why the lunar orbit is always concave to the sun, will be made clear by the following diagram :—

Let the dotted curve A B C D E represent the moon's orbit crossing that of the earth at A, C, and E.   At A, the moon is in first quarter, and *west of the earth* (although *east of the sun*) ; at B, it has made one-fourth of a revolution, and is opposite to the sun and full; at C, it is in last quarter, being east of the earth; at D, it is new; and at E, again west of the earth and in first quarter, having thus completed *one lunation.*   The *arc* A C or C E being known, it it easy to compute the distance of the *chord* A C or C E from the arc.   This will be found to be about 750,000 miles; but as the moon's distance from the earth is only 240,000 miles, its orbit can never be beyond the chord, but must, as at C D E, be within it; and hence, must be always *concave to the sun.*   In the diagram, the *principle* only is illustrated,

Fig. 79.

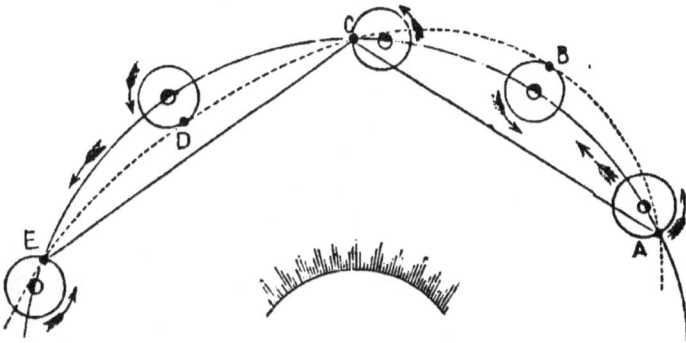

the relative distance of the moon being greatly exaggerated, as well as the orbital movement of the earth during the lunation. The arc A C E in the diagram is more than 120°, whereas it should be only about 29°.

c. **Librations of the Moon.**—As the orbit of the moon is elliptical, her velocity is not uniform, sometimes exceeding that of her rotation, and at other times exceeded by it. In consequence of this, a small portion of the hemisphere turned away from the earth becomes visible alternately at the eastern and western limbs. This is called the *libration* in longitude.* A portion of her surface is also exhibited alternately at each pole, caused by the inclination of her axis to the plane of her orbit. This is called the *libration in latitude.*

The greatest extent of the libration in longitude is 7° 53′; in latitude, 6° 47′; and the whole amount of the moon's surface made visible by both is about $\frac{7}{100}$. There is also a third libration caused by the difference in the angle under which the moon is viewed at any place when on the meridian from that at which we see it when at or near the horizon. This is called the *diurnal libration.* It is, however, quite inconsiderable, amounting to only 32″ when greatest, and bringing into view but $\frac{6}{1000}$ of the moon's surface. Hence, $\frac{576}{1000}$ of the lunar surface is all that we are ever able to see; $\frac{424}{1000}$ having never been gazed at by any human eye.

---

* *Libration* means a *balancing*, and is applied in consequence of the apparent rolling or vibratory motion of the moon from one side to the other.

QUESTIONS.—*c.* What are librations? Of how many kinds? Explain each. How much of the moon's surface have we ever been able to see?

d. **Position of the Lunar Axis.**—The moon's axis leans toward its orbit 6° 39' ; hence, this is the angle which the plane of its equator makes with that of its orbit ; and observation determines that the plane parallel to the ecliptic lies *between these two planes ;* therefore, the inclination of the moon's axis to the plane of the ecliptic is equal to 6° 39'—5° 8', (the inclination of the orbit) ; that is, 1° 31'.

It is a curious fact that the *line of equinoxes* of the moon constantly coincides with the *line of nodes* of its orbit, the ascending node of its orbit being situated at the descending node of its equator. Hence the lunar equinoxes retrograde with the nodes, and the pole of the moon revolves around that of the ecliptic, requiring 18½ years to complete the circuit.

Fig. 80.

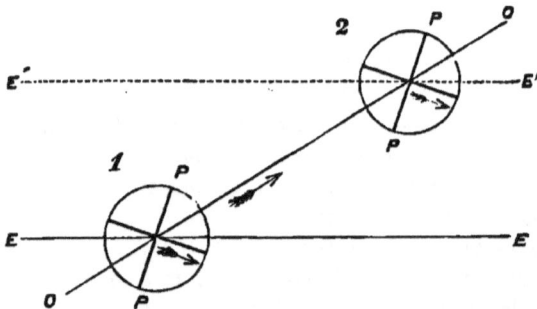

Fig. 80 represents the moon in two positions of her orbit, O O; at 1, in the ascending node, and at 2, when she has her greatest northern latitude. E E represents the plane of the ecliptic, and E' E', a plane parallel to it, each passing between the planes of the moon's equator and orbit, and at the point where the former descends below the latter. The *angular distance* between the planes E E and E' E', of course, never exceeds 5⅓°, which is about ten times the apparent diameter of the moon as seen from the earth. Hence, the greatest distance between these planes is about ten times the diameter of the moon, or 21,600 miles, which at the distance of the sun subtends an angle of about 49'', and to this extent may affect the inclination of the axis to the ecliptic.

174. Owing to the small inclination of the moon's axis to the plane of the ecliptic (1° 31'), she can have but very little change of seasons, and that not constant, because her axis does not always point in the same direction.

a. From what has been said above (Art. 173, *d*), it will be evident that the lunar solstices and equinoxes change places with each other every 9¼ years; whereas, the period required for a similar change in the earth, occasioned by precession, is about 13,000 years.

175. A lunar day must be nearly 15 times as long as one of our days, and a lunar night of the same length; since any place on the moon's surface requires 29½ days to return to the same relative position with the sun. Hence the sun must remain above the horizon during one half of that period, and below it the other half.

a. Mountains situated at either of the lunar poles must have perpetual day; for the sun there can never be more than 1¼° below the horizon; and on so small a body as the moon, the horizon would dip that amount at an elevation of about ⅓ mile.

b. **The Earth's Light.**—On one hemisphere of the moon, the long night must be relieved by the light of the earth, which exhibits the same phases to the moon as the latter does to the earth, except that they are reversed; that is, when the moon is new to us, the earth is full to the moon; and when the lunar form is but a slender crescent, the earth is gibbous, showing itself with almost full splendor. Now, as the earth's disc contains about 14 times as much surface as that of the moon, the light of the earth must cause a very considerable illumination.

c. The effect of this is seen when the moon is just emerging from conjunction, the dark part of her disc being slightly illumined by the light of the nearly full earth, so that the full, round form of the moon's disc becomes visible, the bright crescent appearing at the edge toward the sun. This is sometimes called "the old moon in the new moon's arms."

d. **The Earth appears Stationary to the Moon.**—The earth, although it exhibits phases to the moon, does not appear to revolve around it, but remains at every place on the lunar hemisphere which is turned toward it, nearly at a fixed point in the heavens; this point varying, of course, with the change of place of the observer. This will be obvious, when it is considered that the rotation of the moon would give the earth an apparent motion from *east to west;* but the motion of the moon in her orbit would give it an apparent motion *at the same rate*, from *west to east ;* hence, one counteracts the other, and the earth appears to be almost stationary, only shifting its position backward and forward by the amount of libration.

176. Appearances indicate that the moon has very little, if any, atmosphere; and that its surface is as devoid of water as of air.

a. When viewed with a telescope, the surface of the moon appears entirely unobscured by any clouds or vapors floating over it; and when the moon's edge comes in contact with a star, the latter is immediately extinguished; whereas, if there were an atmosphere, it would, from the effect of refraction, rest on the edge for a short time; that is, it would be visible when a short distance actually behind the moon. Observations of this kind have been made with so much nicety, that it is believed that an atmosphere two thousand times less dense than that of the earth could not have escaped detection. If any atmosphere therefore exists, it must be rarer than the attenuated air in the exhausted receiver of the most perfect air-pump.

b. The absence of water follows from that of air; since, without the latter, the heat of the sun would be incapable of preserving the temperature above the freezing point; as we see on the tops of terrestrial mountains, which are constantly covered with snow, from the extreme rarefaction of the air at those heights. If water existed, it would therefore soon be converted into ice; but we see no indications of it even in this form.

c. Some have accounted for this by supposing that the internal heat of the moon was once very great, as is that of the earth at the present time; but that having cooled, the moon has contracted in volume, and that vast caverns have thus been formed in its interior, into which the water has penetrated, and, of course, disappeared. Indeed, it is obvious that only great internal heat could keep an ocean upon the surface of a body like the earth or moon.

SELENOGRAPHY.

177. That part of the moon's surface which is turned toward the earth has been very carefully observed, and all

the objects upon it delineated upon maps or charts, so as to show their exact forms and relative positions. This branch of astronomical science is called SELENOGRAPHY.*

a. This department of the astronomer's labors has been prosecuted with extraordinary zeal and industry by the Prussian astronomers, Beer and Mädler. Their chart, measuring 37 inches in diameter, exhibits the lunar surface with the most astonishing minuteness and accuracy. Other charts have also been constructed ; and the moon is still receiving a very scrutinizing survey by a number of eminent astronomers, each taking a separate belt or zone, with the object of arriving at still greater minuteness of delineation.

178. The moon's disc when viewed through a telescope presents a diversified appearance of dusky and bright spots; the latter being evidently elevated portions of the surface, and the former, plains or valleys.

a. The dusky patches were once thought to be seas, and they still

Fig. 81.

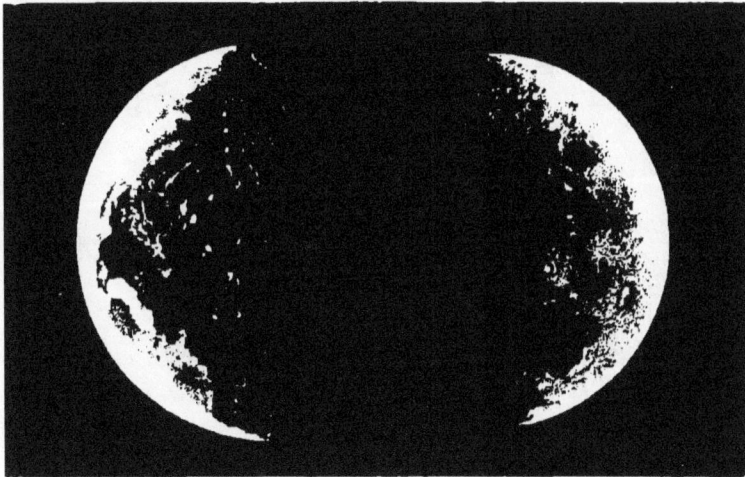

PHOTOGRAPHIC VIEWS OF THE MOON.—*De La Rue.*

* From the Greek word *selenê*, the *moon*, and *graphy*, a description.

QUESTIONS.—*a.* Construction of lunar charts? 178. How does the moon appear when viewed through a telescope ? *a.* What are the dusky patches ?

retain these names in selenography, although without any such literal meaning; thus, one is called *Mare Tranquillitatis*, or Sea of Tranquillity; another, *Mare Nectaris*, Sea of Nectar, etc.

b. **Lunar Mountains.**—Mountains on the moon's surface are indicated by the bright spots that appear scattered over the disc, and beyond the *terminator*, or line that separates the dark from the illuminated part of the disc, and by the shadows cast upon the surface of the moon when the sun shines obliquely upon these elevations.

c. These mountains are of various forms, including with others, the following :—

1. *Rugged and precipitous ranges*, many of a circular form, enclosing great plains, called on this account, "Bulwark Plains," from 40 to

Fig. 82.

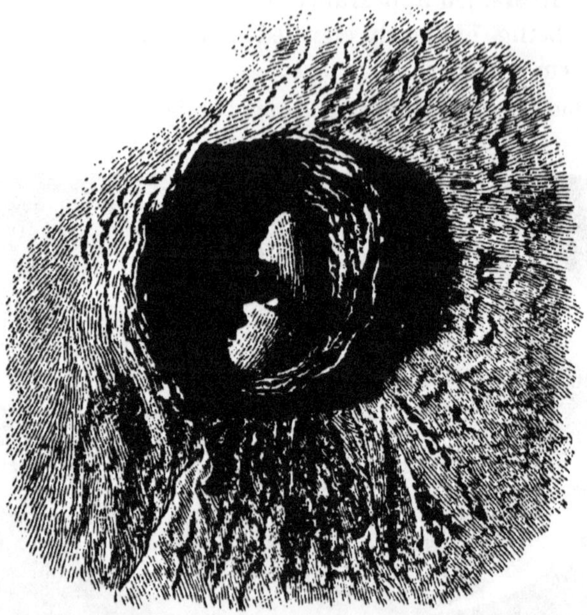

COPERNICUS, FROM A DRAWING BY SIR JOHN HERSCHEL.

120 miles in diameter; 2. *Lofty mountains*, of a circular form, enclosing an area from 10 to 60 miles in diameter, resembling the crater

QUESTIONS.—*b.* How are mountains indicated?   *c.* What classes of mountain formations?

of a volcano but of vast size, and sometimes containing in the centre one or more lofty peaks: such formations are called *Ring Mountains;* 3. Smaller cavities, called *craters,* also enclosing a visible space, and a central mound; and 4. Deep hollows, called *holes,* showing no enclosed area.

From the Ring Mountains, streaks of light and shade radiate on all sides, spreading to a distance of several hundred miles. These are called *radiating streaks.* They are attributed by some to the streams of lava which once flowed in all directions from these evidently volcanic mountains.

d. Copernicus.—(Fig. 82)—This is one of the grandest of the Ring Mountains. It is 56 miles in diameter, and has a central mountain, two of whose six peaks are quite conspicuous. The summit, a narrow ridge, nearly circular, rises 11,000 feet above the bottom. It is very brilliant in the full moon, sometimes resembling a string of pearls. It lies on the terminator a day or two after first quarter. Another of the Ring Mountains (Tycho) is visible to the naked eye, in the southeast quadrant of the moon. It is 54 miles across, and is 16,600 feet high.

e. Height of Lunar Mountains.—Beer and Mädler have calculated the height of more than 1000 mountains, several of which reach an elevation of 23,000 feet, which is nearly equal to that of the loftiest terrestrial peaks; and, of course, relatively very much greater.

Fig. 83.

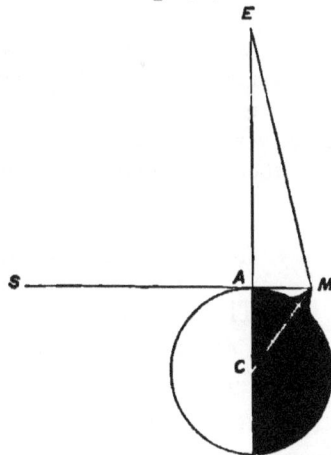

To understand the principle on which the altitude of the lunar mountains is found, let E (Fig. 83) represent the position of the earth, C, the centre of the moon, Ṡ, the direction of a ray of the sun, falling on the top of a mountain at M, which therefore appears to an observer at E, at the distance A M from the terminator at A.. Now, this distance can be found by angular measurement and calculation. Suppose it to be about ₁⁄₂₀ of the apparent diameter of the disc, or about 1'; then A M will be $\frac{1}{30}$ of the moon's diameter, or about 72 miles. Then, from the properties of the right-angled triangle, $(AC)^2 + (AM)^2 = (CM)^2$; that is, $(1080)^2 + (72)^2 = (CM)^2 = 1,171,584$; the square root of which, 1082.4, will be C M. As this is the sum of the moon's radius and the height of the mountain, the latter must be $1082.4 - 1080 = 2.4$ miles.

f. **The General Physical Condition of the Moon's Surface,** therefore, as far as we can observe it, is characterized by uniform desolation and sterility. Sir John Herschel says, that among the lunar mountains is seen in its greatest perfection, the true volcanic character, as observed in the crater of Mt. Vesuvius and elsewhere, except that the internal depth of these lunar craters is sometimes two or three times as great as the external height, and that they are of vastly greater magnitude. By means of the great telescope of Lord Rosse, the interior of some of these craters is seen to be strewed with huge blocks, and the exterior crossed by deep gullies radiating from the centre. No reliable indication of any *active volcano* has ever been obtained ; although, Sir William Herschel, in 1787, asserted that he had seen three lunar volcanoes in actual operation.

g. **Are there People in the Moon ?**—This question has often been discussed, but idly ; since no positive evidence can be adduced on one side or the other. The distance of the moon is too great for us to detect any artificial structures, as buildings, walls, roads, etc., if there were any ; and certainly, without air or water, no animals such as inhabit our own planet could exist there. But the Almighty Creator can place animals and intelligent beings in any part of the universe, and accommodate them to the peculiar circumstances of their abode ; and it would perhaps be strange if He had left even our little satellite without an intelligent witness of His infinite power and beneficence.

IRREGULARITIES OF THE MOON'S MOTIONS.

179. The attraction of the sun acting unequally on the moon in different parts of its orbit gives rise to very many disturbances and irregularities in its motion ; so that it is a very difficult problem to calculate its exact place at any given time.

QUESTIONS.—*f.* Physical condition of the moon's surface ? *g.* Is the moon inhabited ?
179. Lunar irregularities—how caused ?

a. The sun's attracting force upon the moon acts *directly* in conjunction and opposition, but, on account of the difference in the distance, is greater in the former position than in the latter; while at the quadratures, it acts *obliquely,* thus giving rise to a variety of disturbances, or *perturbations.*

b. The attraction of the sun upon the moon is *absolutely* more than twice as great as that of the earth; but being very nearly equal on both earth and moon, they move with regard to each other *almost* as if they were not attracted at all by the sun. If the attraction of the sun upon the earth were suspended, the moon would abandon the earth, and either revolve around the sun, or move directly to it. As the distance of the earth and moon from each other is so small relatively to their distance from the sun (about $\frac{1}{347}$), their mutual attractions are not much disturbed by the action of the sun; but they are *to some extent.* Thus, in conjunction, the moon is attracted more than the earth, but in opposition, less; so that the tendency of the sun's force is to pull them apart when in either of these positions. In the quadratures, however, the sun's force acts obliquely, and consequently tends to pull them together. Hence, we may say, the attraction of the earth upon the moon is *diminished in the syzygies,* and *increased in the quadratures.*

c. The following are the principal irregularities or *inequalities* to which the moon's motion is subject : (Those completed in short periods are called *periodical;* those that require very long periods for their completion are called *secular.*)

1. *Evection,* which is the largest of these inequalities, is the variation in the moon's longitude, due to the action of the sun, above referred to. It depends upon the moon's angular distance from the sun, and the eccentricity of its orbit. By it the equation of the centre of the moon is diminished in syzygies and increased in quadratures. It may influence the moon's longitude to the extent of 1° 20'. This irregularity was discovered by Ptolemy.

2. The *variation,* which also affects the longitude of the moon to the extent of 30'. It arises from the disturbing force of the sun, acting upon the moon when in the octants, or points half-way between the syzygies and quadratures. This was discovered by Tycho Brahe,

and was the first lunar inequality explained by Sir Isaac Newton by applying the law of gravitation.

3. The *annual equation*, which results from the varying velocity of the earth in its orbit. It may affect the moon's longitude about 11'.

4. The *parallactic inequality*, arising from variations in the disturbing force of the sun upon the moon according as the latter is in that part of its orbit nearest to, or farthest from, the sun. It may affect the moon's longitude to the extent of 2'.

5. The *secular acceleration* of the moon's mean motion, caused by the diminution of the eccentricity of the earth's orbit. At present it amounts to 10″ every 100 years, the periodic time of the moon being constantly diminished to that extent. This was discovered by Halley in 1693, by comparing the periodic time of the moon, as deduced from Chaldean observations of eclipses made at Babylon, 720 and 719 B.C., with Arabian observations made in the 8th and 9th centuries A.D. La Place demonstrated its cause. At a very distant period, this inequality will, of course, be reversed, becoming a *retardation* instead of an *acceleration*.

d. Other irregularities have been discovered, caused by the disturbing action of Venus. These various inequalities constitute what is called the *Lunar Theory ;* and when they are all applied, the *computed* place of the moon should precisely agree with the *observed* place.

QUESTIONS.—The annual equation ? The parallactic inequality ? The secular acceleration ? *d.* What other inequalities ? The Lunar Theory ?

CHAPTER X.

ECLIPSES.

180. An ECLIPSE * is the concealment or obscuration of the disc of the sun or moon by an interception of the sun's rays. Eclipses are, therefore, either *Solar* or *Lunar*.

181. A SOLAR ECLIPSE is caused by the passage of the moon between the earth and sun so as to conceal the sun from our view.

182. A LUNAR ECLIPSE is caused by the passage of the moon through the earth's shadow.

a. By a shadow is meant simply the space from which the light of a luminous body is wholly intercepted by the interposition of some *opaque* body. Since light proceeds from a luminous body in straight lines, and in all directions, the darkened space formed behind the earth or moon must be *conical ;* that is, of the form of a cone, circular at the base and terminating at a point; since the sun or luminous body is larger than either of the opaque bodies. The shadow is sometimes called by its Latin name, *umbra.*

b. Besides the totally darkened space called the *umbra*, there is formed on each side a space from which the light is only partially excluded; this is called the *penumbra.*† The relations of the umbra

---

* From the Greek word *ekleipsis*, which means a *fainting away*. The ecliptic is so called because eclipses only take place when the moon is in its plane.

† From the Latin word *pene*, meaning *almost*, and *umbra*, meaning *a shadow.*

QUESTIONS.—180. What is an eclipse? Of how many kinds? .181. How is a solar eclipse caused? 182. A lunar eclipse? *a.* How is a shadow defined? The form of the earth's or moon's shadow? *b.* Define the terms *umbra* and *penumbra.*

to the penumbra will be understood by inspecting the annexed diagram (Fig. 84).

Fig. 84.

SOLAR AND LUNAR ECLIPSES.

183. If the moon moved exactly in the plane of the
earth's orbit, a solar eclipse would occur at every new moon,
and a lunar eclipse at every full moon; but as the moon's
orbit is inclined to that of the earth, an eclipse can only
happen when the moon is at or near one of its nodes.

a. When the moon is new or full at a considerable distance from
its node, it is too far above or too far below the plane of the ecliptic to
intercept the sun's rays from the earth, or to pass within the limits of
the earth's shadow. It will be easily understood that no eclipse can
occur unless the sun, earth, and moon are situated exactly or nearly
in the *same straight line.* [See Fig. 84.]

b. The limit north or south of the ecliptic within which an eclipse
must occur is larger in the case of solar than in the case of lunar
eclipses. In the former it varies from 1° 35' to 1° 24'; in the latter,
from 63' to 52'.

QUESTIONS.—183. Where must the moon be when an eclipse occurs? *a.*   How explained? *b.* What is the limit in latitude for solar and lunar eclipses? Explain and
demonstrate each by Fig. 85.

Fig. 85.

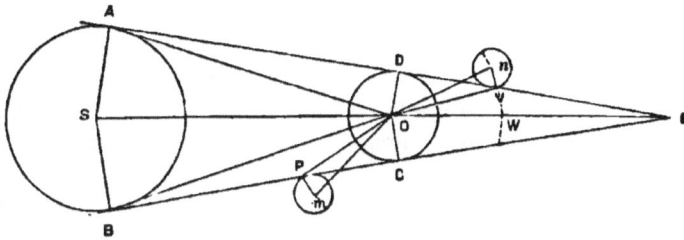

To explain how this is found, let S be the centre of the sun, and O the centre of the earth, S O E being the plane of the ecliptic; let also P be the position of the moon at the limit for a solar eclipse, and V, its position for a lunar eclipse. The angular distance of the *moon's centre* from the ecliptic in each case is the limit required; S O *m* is that angle for the former, and *n* O E, for the latter. Now, S O *m* is equal to S O B+P O *m*+ B O P; and S O B is the apparent semi-diameter of the sun, and P O *m* is that of the moon. But, O P C, being exterior to the *triangle* B O P, is equal to the sum of the two interior angles O B P and B O P, and hence B O P is equal to O P C—O B C, or the moon's horizontal parallax *minus* that of the sun. Therefore, the solar limit in latitude is equal to the *sum of the apparent semi-diameters of the sun and moon increased by the difference between the horizontal parallax of each ;* or 16½′ + 16½′ + 1° 2′ = 1° 35′, when greatest (omitting the sun's parallax, which is very small); and 15½′ + 14½′ + 53½′ = 1° 24′, when least. Hence, when the moon's latitude at the time of inferior conjunction does not exceed the former, an eclipse *may* occur; when it does not exceed the latter, an eclipse *must* occur.

The angle of limit for a lunar eclipse is *n* O E, obviously less than S O *m*. It is composed of the angle *n* O V, or the apparent semi-diameter of the moon, and the angle V O E, or the angle subtended by one-half of the diameter of the shadow, where the moon traverses it. Now, V O E = O V D —O E V, and O E V = A O S—O A D; hence V O E = O V D+O A D —A O S. But O V D is the moon's horizontal parallax, O A D is that of the sun, and A O S is the sun's apparent semi-diameter. Consequently, *n* O E, or the lunar limit in latitude, is equal to the *sum of the horizontal parallax of the sun and moon, diminished by the sun's apparent semi-diameter, and increased by that of the moon.* That is, 62′ — 15½′ + 16½′ = 63′, when greatest ; and 53½′ — 16½′ +14½′ = 52′, when least. These calculations, being made only for illustration, are but approximatively correct.

184. The distance in longitude, either side of the node,

ECLIPSES.

within which an eclipse can occur, is called the ECLIPTIC LIMIT.

185. The solar ecliptic limit extends about 17° on each side of the node; the lunar ecliptic limit, about 12°.

a. This difference follows from the difference in the limits in latitude, the ecliptic limits in longitude being computed from those in latitude.

Fig. 86.

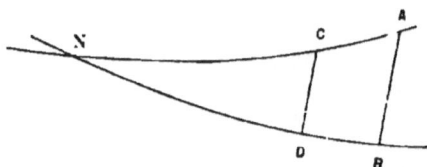

For in Fig. 86, let B N be a portion of the ecliptic, A N, a part of the moon's orbit, N, the node, A B, the solar limit in latitude and C D, the lunar. It will be at once apparent that since A B is greater than C D, it must be farther from the node. To calculate the exact amount, there are given, in the right-angled triangle A N B, the angle at $N = 5\frac{1}{2}°$; the side A B or C D, and the right angle at B, to find the side B N or D N, which can easily be done by the higher mathematics.

b. Since the limits in latitude vary, those in longitude also vary, the amount given above being the mean. The greatest solar ecliptic limit is 18° 36′; and the least, 15° 20′; the greatest lunar ecliptic limit is 12° 24′; and the least, 9° 23′. Within the former, an eclipse may happen; within the latter, it must.

Fig. 87.

ASCENDING NODE

PATH OF THE SUN CROSSED BY THAT OF THE MOON.

Fig. 87 illustrates the relative position of the sun and moon's orbits, with respect to the ecliptic limits. In the centre the moon is exactly at the ascending node; while at the extremes, it is at the limits both in latitude and longitude. Except at the node, the moon, it will be apparent, only partially covers the disc of the sun, within the limits on each side.

QUESTIONS.—185. What is the extent of the solar and lunar ecliptic limits? *a.* Why do they differ? *b.* Why is each not always the same?

ECLIPSES.

186. Since the solar ecliptic limits are wider than the lunar, eclipses of the sun are more frequent than those of the moon.

187. The *greatest number* of eclipses that can happen in a year is seven; five of the sun and two of the moon, or four of the sun and three of the moon. The *least number* is two, both of which must be of the sun.

a. The usual number is four, and it is rare to have more than six. From the above statement, it will be seen that the greatest number of solar eclipses is five, and the least two; and that the greatest number of lunar eclipses is three, while none at all may occur during the year.

b. Number of Solar Eclipses.—Since the sun crosses the line of nodes twice each year, and his monthly progress in the ecliptic is about 29°, while a solar eclipse must occur if the moon is within 15° 20' of either node, or within a space of 30° 40', there must evidently be a solar eclipse each time the sun passes the node, or twice each year. Now, if the sun, at the time of new moon is 18° west of the node, it *may* be eclipsed (Art. 185, *b*); and if it were, there would be another eclipse at the next new moon, for the sun would have advanced less than 11° east of the node. Again, in six lunations from the first new moon referred to, the sun would have advanced 174°, and consequently would be 174°—18°, or 156° east of one node, and 24° west of the other; but the node is itself moving to the west about 1½° every lunation; and hence, the sun would be only 24°—9°, or 15°, from the node, so that a third eclipse would take place; and after another lunation, a fourth, since the sun would then be less than 15° from the node. Now, owing to the retrogradation of the nodes, the sun passes from one to the same again in 346 days; and hence, if it passed one at the beginning of the year, it would pass it again toward the end of the year, and there would be three passages of a node in that time; so that if four eclipses had previously taken place, there might be still another toward the end of the year, making *five* in all.

c. Number of Lunar Eclipses.—As the space on each side of the node, within which a lunar eclipse *must* occur, is only about 9½°, or 19°

on both sides, it is obvious that there might be no lunar eclipse during the year; but, since an eclipse *may* occur within a space of 25° (12° 24' on each side of the node), it follows that one lunar eclipse may occur at each passage of the sun, or three during the year. But three lunar eclipses can not be preceded by five solar eclipses in the same year; for two solar eclipses can not take place at each node, unless, at the first one, the sun is at least about 15° west of the node, so that there would not be enough space at the end of the year for both a solar and a lunar eclipse.

188. Solar eclipses do not actually occur as often as lunar eclipses at *any particular place ;* because the latter are always visible to an entire hemisphere, whereas the former are only visible to that part of the earth's surface covered by the moon's shadow or its penumbra.

a. That the moon, in a lunar eclipse, is concealed from an entire hemisphere, will be obvious from the fact that the diameter of the earth's shadow where the moon crosses it is always more than twice as great as the diameter of the moon, and is sometimes nearly three times as great. For the angle V O W (Fig. 85) is equal to the sum of the horizontal parallax of the sun and moon, diminished by the apparent semi-diameter of the sun (183, *b*). The greatest parallax of the moon is about 62', and the least, 53¼'; and the least apparent semi-diameter of the the sun is 15¾', and the greatest, 16¼'; hence, the angle V O W is, when greatest, 44¼' (omitting the sun's parallax); and when least, 37¼'; the mean being about 40¾'. As this is the angular value of the semi-diameter of the shadow, it must be doubled for the whole, which therefore is, when greatest, 88¼'; least, 74¼'; mean, 81¼'. Hence, as the moon's apparent diameter is, when greatest, 33¼'; least, 29½'; mean, 31½', the truth of the above statement will be apparent.

b. **Length of the Earth's Shadow.**—This can be readily found by comparing the triangles A S E and D E O (Fig. 85), which being both right-angled triangles, and having all their angles respectively equal, have, by a principle of geometry, proportional sides; so that A S : D O : : S E : O E. But D O, the semi-diameter of the earth, is about $\frac{1}{107}$ of A S, the semi-diameter of the sun; hence O E, the length of the shadow, must be $\frac{1}{107}$ of S E; and therefore, S O must be $\frac{108}{107}$ of the whole distance S E, and, of course, S E, $\frac{107}{108}$ of S O; but O E, the length of the shadow, is equal to S E — S O; hence it is equal to $\frac{1}{108}$ of S O, or the distance of the earth from the sun. Therefore its greatest length is about 877,000 miles, and its least, 850,000 miles; that is, at its mean length, a little more than the diameter of the sun.

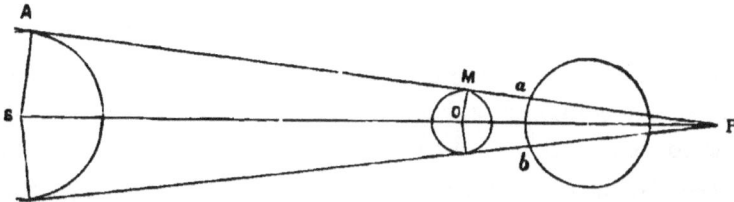

Fig. 88.

c. **Length of the Moon's Shadow.**—This can be found by a similar calculation. Let S (Fig. 88) be the centre of the sun, and O, that of the moon ; P will then be the end of the shadow, and O P its length ; and, in the triangles A S P and O M P, A S : M O :: S P : O P. Now, M O is about $\frac{1}{397}$ of A S ; hence O P is $\frac{1}{397}$ of S P, and S O, $\frac{396}{397}$ of S P; or S P, $\frac{397}{396}$ of S O; therefore O P is $\frac{1}{396}$ of S O, *the distance of the moon from the sun.* Now, the moon's distance from the earth varies between 252,000 miles and 226,000 miles ; and the earth's distance from the sun, between 93 millions and 90 millions ; hence, at the mean distance of the earth and moon, the length of the shadow is about 232,000 miles, or 6,000 miles from the earth's centre, and 2,000 miles from its surface. When the earth is in aphelion and the moon in perigee, it extends about 10,000 miles beyond the earth's centre, or 14,000 miles from the surface a b, which is the *maximum.* When the earth is in perihelion and the moon in apogee, the shadow is about 228,000 miles long, while the moon is 252,000 miles from the earth's centre ; so that it fails to reach the surface of the earth by 20,000 miles.

d. **Breadth of the Moon's Shadow.**—When the end of the shadow extends to the greatest distance beyond the earth's centre, the amount of surface covered by it is the greatest possible. Let a b (Fig. 88) be the diameter of the shadow where it intersects the earth ; and since it is a *very small arc*, we may find its approximate length by considering it a straight line. We shall then have, by comparing the triangles, $A P : a P :: A S : \frac{1}{2} a b$; but a P is 14,000 miles (183 c), and A P is 93,014,000 miles. Hence, $93,014,000 : 14,000 :: 426,000 : \frac{1}{2} a b = 64$ miles+. Therefore $a b$, or the breadth of the shadow where it intersects the earth is about 128 miles. *The breadth of the portion of the earth's surface covered by the shadow is, really,* 1° 54', or 130 miles. This is the maximum. The breadth of the greatest portion of the earth's surface ever covered by the moon's *penumbra* is 70° 17', or 4,850 miles.

189. When the whole of the sun's or moon's disc is concealed, the eclipse is said to be *total;* when only a part of it is concealed, it is said to be *partial.*

190. In order to measure the extent of the eclipse, the apparent diameters of the sun and moon are divided into twelve equal parts, called *digits.*

Fig. 89.

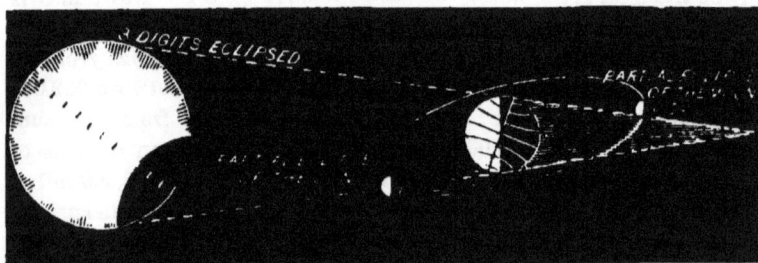

A PARTIAL ECLIPSE OF THE SUN AND MOON.

a. The conditions of a total and a partial eclipse will be apparent from the explanations already given. When the centres of the sun and moon coincide, that is, when the latter is exactly at the node, the eclipse is said to be *central.* A central eclipse of the moon must, of course, be total ; but a solar eclipse may be central without being total ; since sometimes, as it has been demonstrated, the shadow of the moon does not reach the earth. The moon, when this is the case, covers

<hr/>

only the central part of the sun's disc, leaving a ring of luminous sur-face visible around the opaque body. This is called an *annular* *
eclipse.

191. An ANNULAR ECLIPSE is an eclipse of the sun, which happens when the moon is too far from the earth to conceal the whole of the sun's disc, leaving a bright ring around the dark body of the moon.

192. The time at which an eclipse will occur may be dis-covered by finding the mean longitudes of the sun and node at each new or full moon throughout the year, and compar-ing the difference of the longitudes with the ecliptic limits.

Fig. 90.

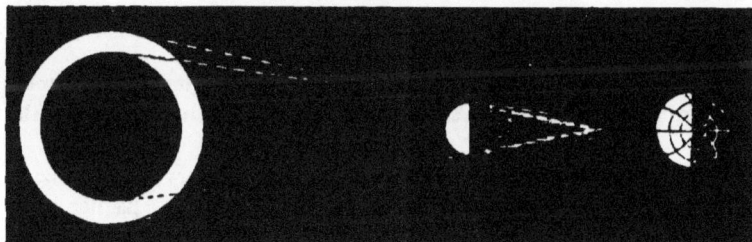

AN ANNULAR ECLIPSE.

a. Cycle of Eclipses.—Eclipses of both the sun and moon recur in nearly the same order, and at the same intervals, after the expiration of 18 years and 10 or 11 days (according as there may be 5 or 4 leap-years in this period). For a lunation is about 29.53 days, and the time of a revolution of the sun with respect to the node, 346.62 days, which periods are nearly in the ratio of 19 to 223 ; so that 223 lunations are almost equal to 19 revolutions of the sun ; and 346.62 days $\times$ 19 $=$ 18^y 11$\frac{1}{4}$^{d}. This is called the *cycle or period of eclipses.* The eclipses which occur during one such period being noted, subsequent eclipses may easily be predicted ; as their order is the same, only they are 10 or 11 days later in the month, and about eight hours later in the day ; so

* From the Latin word *annulus*, meaning a *ring.*

that in one cycle eclipses may be visible, and in the next invisible, to a particular place. During this period there are generally 41 solar and 29 lunar eclipses. This cycle was known to the ancient Egyptians and Chaldeans, and called by them *Saros*.

193. The phenomena connected with a total eclipse of the sun are of a peculiarly interesting character, and have been observed by astronomers with great attention and industry.

a. To an ignorant mind, this occurrence must be the occasion of very great awe, if not actual terror. A universal gloom overspreads the face of the earth as the great luminary of day appears to be expiring in the sky ; the stars and planets become visible, and the animal creation give signs of terror at the dismal and alarming aspect of nature. Armies about to engage in battle have thrown down their arms and fled in dismay from the seeming anger of heaven. This was the case at the eclipse predicted by Thales, which occurred on the eve of the battle between the Medes and Lydians, 584 B.C.

b. **Phenomena of a Solar Eclipse.**—The following are the most interesting of the phenomena presented during a total eclipse of the sun :—

1. *The change of color in the sky* from its ordinary blue or azure tinge to a dusky, livid color intermixed with purple. Kepler mentions that during the solar eclipse in 1590, the reapers in Styria noticed that every thing had a yellowish tinge. The darkness is not, however, total, but sufficiently great to prevent persons' reading.

2. The *corona*, or halo of light which appears to surround the moon while it covers the disc of the sun. This is, at the present time, supposed to be caused by the atmosphere of the sun.

3. When the moon has almost covered the disc of the sun, leaving only a line of light at the edge, this line is broken up into small portions, so as to appear like a band of brilliant points. This phenomenon is called *Baily's beads*, from Mr. Francis Baily, who was the first to describe it minutely, in 1836. This is supposed to be caused by the irregularities of the moon's surface, serrating its dark edge, and projected on the sun's brilliant disc.

4. *Pink or rose-colored protuberances* which project from the margin of the moon's disc when the obscuration is total. One measured by

---

QUESTIONS.—193. The phenomena connected with a total eclipse ? *a.* Effect on ignorant minds ? *b.* State the most interesting phenomena presented by a solar eclipse ? The corona ? Baily's beads ? Rose-colored protuberances ? Explain the cause of each.

De La Rue, in 1860, was found to be at least 44,000 miles in vertical height above the sun's surface. They have been seen by most observers. No entirely satisfactory cause has been assigned for these appearances, although it seems to be settled that their origin is in the sun and not the moon ; and it is thought by some that they are clouds floating in the atmosphere of the sun, their peculiar color being caused by the absorption of the other colors, as sometimes occurs in the case of clouds in our own atmosphere.

c. **Appearance of a Lunar Eclipse.**—In a total lunar eclipse, the moon does not become wholly invisible, but assumes a dull, reddish hue, which arises from the refraction of the sun's rays by the earth's atmosphere. The red color is caused by the absorption of the blue rays in passing through the atmosphere, just as the western sky assumes a ruddy hue when illuminated in the evening by the solar light. Sometimes, however, it happens that the moon is rendered very nearly invisible, as was the case in 1642 and 1816 ; and the degree of distinctness of the moon's appearance varies considerably at different times, owing to the different conditions of the atmosphere.

d. **Earliest Observations of Lunar Eclipses.**—These were made by the Chaldeans,—the first recorded eclipse having taken place in 720 B.C. This eclipse was total at Babylon, and occurred about 9½ o'clock P.M. The record of the occurrence of eclipses is often very useful in fixing the dates of history.

194. An OCCULTATION is the concealment of a planet or star by the interposition of the moon or some other body.

The occultation of a planet or star by the moon is a very interesting and beautiful phenomenon. From new moon to full moon, she advances eastward with the dark edge foremost, so that the occulted body disappears at the dark edge and re-appears at the enlightened edge. In the other part of her orbit this is reversed. The former phenomenon is of course the more striking, the star or planet appearing to be extinguished of itself.

CHAPTER XI.

THE TIDES.

195. TIDES are the alternate rising and falling of the water in the ocean, bays, rivers, etc., occurring twice in about twenty-five hours.

196. FLOOD TIDE is the rising of the water, and at its highest point is called *high water*. EBB TIDE is the falling of the water, and at its lowest point is called *low water*.

197. The tides are caused by the unequal attraction of the sun and moon upon the opposite sides of the earth.

a. Since the attraction of gravitation varies inversely as the square of the distance, the sun and moon must attract the water on the side nearest to them more than the solid mass of the earth; while on the side farthest from them, the water must be attracted less than the solid earth; hence, there must be a tendency in the water to rise at each of these points; it being drawn away from the earth at the point toward the sun or moon, and the earth being drawn away from it at the other point. At the points 90° from these, the effect of the attraction is just the reverse; for since it does not act in parallel lines, it tends to draw together the two sides of the earth, and thus compresses the water so as to cause it to recede, and hence increases its tendency to rise at the other points.

Thus at A (Fig. 91) the water is attracted by the sun more than the earth, and at B less; while at C and D the attraction squeezes in the water, as it were, so as to make it recede still more, and thus to augment its rising at A and B. It will be obvious that the attraction of the moon acts so as to disturb the water just as that of the sun does; hence, in the position repre-

sentcd in the diagram, when the moon is in opposition, the action of both the
sun and moon is exerted upon the same points, A, B and C, D; and it will
also be obvious that if the moon were in conjunction, that is, on the same
side of the earth as the sun, the effect would be the same, because the
same points would be acted upon.

Fig. 91.

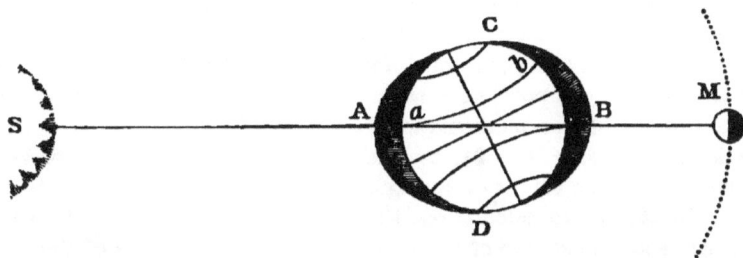

b. From the above explanation, it will be apparent that similar
tides must occur simultaneously on opposite sides of the earth ; namely,
flood tides on the side turned toward and that turned from the sun or
moon, and ebb tides at the two points 90° distant.

198. Since the moon is so much nearer to the earth than
the sun is, its attraction on the opposite sides is much more
unequal, and consequently its disturbing action is greater.
At the mean distance of the sun and moon from the earth,
the disturbing or tidal force of the moon is about $2\frac{1}{4}$ times
that of the sun.

a. Considering the moon's mean distance from the *centre* of the
earth 238,000 miles, and the earth's diameter 8,000 miles, the moon
must attract the side of the earth nearest to her more than the opposite side in the ratio of $(234,000)^2$ to $(242,000)^2$: that is, as 1 to 1.07
(nearly). Hence, the moon's disturbing force is .07 of her own attraction. Taking the sun's mean distance from the earth's centre at
91,500,000 miles, the ratio of his different attractions will be as
$(91,496,000)^2$ to $(91,504,000)^2$, or as 1 to 1.000175 (nearly) ; that is, the
sun's disturbing force is equal to .000175 of his own attraction.

Now, the mass of the sun is about 25,200,000 times that of the moon [315,000 × 80], and its distance, 385 times as great ; hence, $\dfrac{25,200,000}{(385)^2} =$ 170 (nearly) will be the force of the sun upon the earth, the moon's being 1 ; in other words, the sun's attraction on the earth is to the moon's as 170 to 1. Hence, 170 × .000175, which is equal to .02975, is the sun's disturbing force, while the moon's, as above shown, is .07 ; and therefore the former is to the latter as .02975 to .07, or as 1 to $2\frac{1}{2}$. This calculation being designed merely to illustrate the principle, is made with only approximate accuracy, but gives the true result as found by the higher mathematics. Sir Isaac Newton computed this ratio as 23 to 58, or 1 to $2\frac{1}{2}$, by calculations based upon the observed differences of the height of spring and neap tides.

b. In the above calculation, the moon's mass is supposed to be known, and is made use of to determine the relative forces of the sun and moon ; but, practically, the problem is reversed, the comparative disturbance of the two bodies being deduced by observations of the tides themselves, and then employed to determine the moon's mass.

199. When the sun and moon are on the same or opposite sides of the earth, they unite their attractions, and thus raise the highest flood tides at the points under or opposite them and the lowest ebb tides at the points 90° from these. Such tides are called *spring tides ;* and they occur at every new and full moon, or a short time afterward.

200. When the moon is in quadrature, its tidal force is partly counteracted by that of the sun, since the two forces act at right angles with each other; and consequently the water neither rises so high at flood, nor descends so low at ebb tide. Such tides are called *neap tides ;* they occur when the moon is in either of the quarters.

Fig. 92 represents neap tide. The effect of the sun at A and B, and of the moon at C and D, is to equalize the height of the water all over the earth. The pupil must understand in inspecting these diagrams, that the actual effect of the sun or moon is not, by any means, so great as is represented. It is, in fact, but a few feet, while the earth's diameter is nearly 8,000 miles.

Fig. 92.

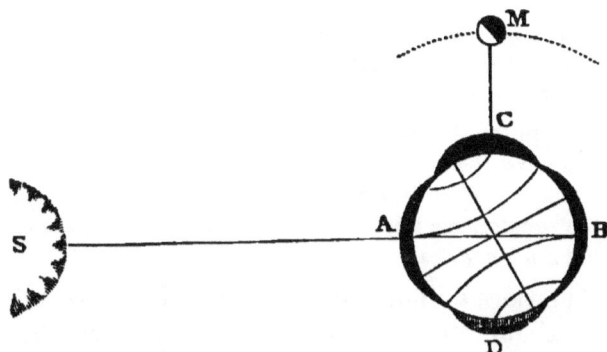

a. Since the tidal force of the moon is so much greater than that of the sun, it is the passage of the former across the meridian that determines the rising of the tide at any place, this lunar tide being either augmented or diminished by the inferior tidal force of the sun.

b. The tides not only vary according to the position of the moon with regard to the sun, but are sensibly affected by the variations in the distance of the moon from the earth, increasing and diminishing *inversely* with it, but in a more rapid ratio.

c. The height of the spring tides is to that of the neap tides, generally, as $2\frac{1}{4}$ to 1. For spring tide is the result of the *sum* of the moon and sun's forces, or $2\frac{1}{4} + 1 = 3\frac{1}{4}$; and neap tide, the result of the *difference*, or $2\frac{1}{4} - 1 = 1\frac{1}{4}$; and $3\frac{1}{4} : 1\frac{1}{4} :: 2\frac{1}{4} : 1$. This varies at different places; at Brest, in France, the spring tides rise to the height of over 19 feet, and the neap tides about 9 feet. On the Atlantic coast of the United States, the height of the former is to that of the latter as 3 to 2.

d. In the northern hemisphere, the highest tides occur during the day in summer, and during the night in winter; but in the southern hemisphere this is reversed.

This will be apparent from an inspection of Fig. 91. The greatest tidal

QUESTIONS.—*a.* What determines the rising of the tide at any place? *b.* What additional cause of variation in the tides? *c.* How does the height of spring tides compare with that of neap tides? *d.* What difference between the diurnal and nocturnal tides? Why?

elevation of the water is of course at A and B, and diminishes north and south of these points. At *a* the elevation of the water is obviously greater than at *b;* but *a* is the position of a place in the northern hemisphere, at noon, and in summer, since the axis is turned toward the sun, and *b*, its position at midnight; so that the tide is higher during the day than during the night, in this season. Conceive the axis turned the other way, and it will be at once seen that the reverse is true in winter.

e. The height of the tide, therefore, varies with the declinations of the sun and moon, being greater in proportion as the two bodies are near the equator. If at the time of the equinoxes the moon happens to be near the equator, the tides are the highest of all, and are called the *Equinoctial Spring Tides.*

201. The tides do not rise at the same hour every day, but generally about 50 minutes later; because, as the moon advances in her orbit, the same place on the earth's surface does not come again under the moon until about 50 minutes later than on the previous day.

a. The interval which elapses from the moon's passing the meridian of a place until it returns to the same again is 24ʰ 50ᵐ 28ˢ; the interval, therefore, between two successive tides is 12ʰ 25ᵐ 14ˢ. This is not, however, always the true interval, from circumstances which will be explained hereafter.

202. The tide does not generally rise until two or three hours after the moon has passed the meridian; because, on account of its inertia, the water does not immediately yield to the action of the sun or moon.

a. By *inertia* is meant the resistance which matter of every kind makes to a change of state, whether of rest or motion; that is, it can not put itself in motion, neither can it stop itself. The tides are not only retarded by inertia, but, to some extent, by the friction on the bed of the ocean or the sea, or the sides of rivers and confined channels.

b. Solar and Lunar Tide Waves.—The solar tide wave is more retarded than the lunar, since the tidal force of the sun is so much feebler than that of the moon. The general retardation of the tides depends on the *relative position of the solar and lunar tide waves.* At new and full moon the solar tide wave, being more retarded, is east of the lunar; and, therefore, high water, which results from the union of the two waves, must be east of the place it would have been if the moon had acted alone; and hence, on this account, the tide will rise later. When, however, the moon is in either of the quarters, the solar tide wave is west of the lunar, and the tide rises earlier. This is called the *priming and lagging of the tides;* since it either shortens or lengthens the tidal day of 24^h 50^m 28^s. The highest spring tide rises when the moon passes the meridian about $1\frac{1}{2}^h$ after the sun; for then the two tide waves immediately coincide.

c. These tide waves are not to be conceived as currents moving progressively through the ocean, but as undulations rising nearly under the sun and moon, and, as the earth turns on its axis, moving westward over its surface, at the same rate. This would be the case exactly, and at all parts of the earth, if it were uniformly covered with water, so that the great tide wave could move without any obstruction from opposing shores. In the open ocean it constantly follows the moon at the distance of about $30°$ from her; but the tide rises at every place at a different time owing to the peculiarities of its situation. Lines drawn on the map or globe through all the adjacent places which have high water at the same time, are called *cotidal lines.*

d. The average interval of time between noon and the time of high water at any port on the days of new and full moon, is called the *establishment of the port.* The *mean establishment* of New York is about $8\frac{1}{4}$ hours; of Boston, $11\frac{1}{2}$ hours, of San Francisco, $12\frac{1}{4}$ hours.

203. The tides that occur in rivers, narrow bays, or other bodies of water at a distance from the ocean, are not caused by the immediate action of the sun and moon, but arise from the undulations of the great ocean tide wave, urging the water into these contracted inlets. The tides of the ocean are called *primitive tides;* those of rivers, inlets, etc., are called *derivative tides.*

204. The average height of the tide for the whole globe is about 2½ feet; and this is the height to which it rises in the ocean. The height, however, at any particular place depends upon its situation; the highest tides occurring in narrow bays, and arms of the sea running up into the land. Lakes have no perceptible tides.

a. The highest tides in the world take place in the Bay of Fundy, the mouth of which is exposed to the great Atlantic tide wave. At the head of the bay the ordinary spring tides rise to the height of 50 feet, while special tides have been known to rise as high as 70 feet. In New York, the height of the tide is, at its maximum, about 8 feet; in Boston, 11 feet. On the other hand, at some places there is scarcely any tide at all. An instance of this is found at a point on the south-eastern coast of Ireland, the tide stream being diverted to the opposite shore by a promontory at the entrance of St. George's Channel.

b. **Velocity of the Tide-Wave.**—This is affected by various circumstances; such as the depth of the water, the obstructions from opposing shores, etc. The moon tends to draw the water along with it at the rate of 1,000 miles an hour at the equator; but the actual rate of movement is much less rapid. The tide wave requires about 40 hours to reach the Atlantic coast from the south-eastern Pacific, where it originates, traversing the Pacific, Indian, and South Atlantic oceans. When it strikes the shallow waters of the coast its velocity is greatly diminished, not exceeding, sometimes, 50 miles an hour. Its breadth is, of course, diminished with its velocity. At a velocity of 600 miles an hour, its breadth would be over 7,000 miles; but when the velocity is reduced to 100 miles an hour, its breadth is only about 1200 miles.

c. **Atmospheric Tides.**—The same causes that act to disturb the ocean must also produce similar disturbances in the atmosphere. The atmospheric tides, however, have been demonstrated by Laplace to be very inconsiderable in height, not exceeding, at Paris, one-thousandth of an inch,—an amount far too small to be indicated by ordinary observations with the barometer.

QUESTIONS.—204. What is the average height of the water for the whole globe? On what does the height at particular places depend? *a.* Where are the highest tides? Why? *b.* Velocity of the tide wave? *c.* Atmospheric tides?

CHAPTER XII.

I. MERCURY. ☿

205. MERCURY is remarkable for its small size, its swift motion, and the great inclination and eccentricity of its orbit. It is, as far as is positively known, the nearest planet to the sun.

a. **Name and Sign.**—This planet probably derived its name from the swiftness of its motion, Mercury being, in the heathen mythology, the "messenger of the gods." The sign ☿ is supposed to represent the *caduceus*, or wand, which the god is always seen, in the pictures of him, to carry in his hand.

b. **Vulcan.**—Reference is made in Art. 16 to a planet supposed by some to exist between Mercury and the sun ; the following are the circumstances connected with its supposed discovery :—On the 26th of March, 1859, a small dark body was seen to pass over a portion of the sun's disc, by M. Lescarbault, a French physician, but an amateur of astronomy ; and this appeared to him to indicate the existence of a planet whose orbit must be included within that of Mercury. From the observations which he made with his rude instruments, he calculated its period at about 20 days ; its distance, 14,000,000 miles ; and the inclination of its orbit, about 12°. On publishing this fact, the celebrated French astronomer and mathematician, Leverrier, visited him, and after closely questioning him as to his means and method of observation, was completely satisfied of the truth of his statements Singular to say, however, no other observer has been able to detect any indications of such a planet ; but, on the contrary, M. Liais, an

---

QUESTIONS.—205. For what is Mercury remarkable? *a.* Its name and sign? *b.* Supposed discovery of Vulcan?

astronomer of skill and experience, who happened to be engaged in observations of the sun, at Rio Janeiro, at the identical moment of M. Lescarbault's alleged discovery, asserted positively that no planetary object was visible at that time. The existence of any planet inferior to Mercury is therefore considered very doubtful.

206. Mercury and Venus are known to be inferior planets, 1. Because their greatest elongation is always less than 90° ; 2. Because they exhibit all the different phases which are presented by the moon ; and, 3. Because they are seen, at the time of a transit, to pass across the sun's disc.

a. A superior planet also exhibits phases, but it must always show more than half the disc , that is, it must present the full or gibbous form. An inferior planet, however, in passing between the points of extreme elongation, presents the crescent form, and, in inferior conjunction, either totally disappears or is projected, in the form of a small round black spot, upon the disc of the sun.

207. The greatest angular distance of an inferior planet from the sun, during any single revolution, is called its *extreme elongation.* The greatest extreme elongation of Mercury is 28¼° ; the least, 18°.

a. This large variation in the extreme elongation is an indication that Mercury revolves in an orbit of considerable ellipticity, since this angle depends upon the relative distances of Mercury and the earth from the sun. It must be greatest when the earth is in perihelion and Mercury in aphelion, and least when the earth is in aphelion and Mercury in perihelion ; while the mean distance of each would give the mean value of this element.

In Fig 93 let S be the sun, E, the earth, and M, Mercury at the point of extreme elongation, M E being tangent to the orbit, and S M E a right angle. It will be obvious that M E S, the angle of extreme elongation, will be at its maximum when S E is the shortest and S M the longest, and at its minimum when these are reversed ; because its size depends upon the ratio of S M to S E, being greatest when the ratio is greatest. The perihelion distance of the earth is about 90,000,000 miles, the aphelion distance of Mercury is about 42,600,000 miles. For these values the ratio of S M to S E would be about .473; and the angle corresponding to this is 28° 15'. The least ratio of these lines is .3, and hence, the least angle, 18°; while the mean ratio is .387 (nearly), indicating an angle of 22° 47', which is therefore the mean value of this element.

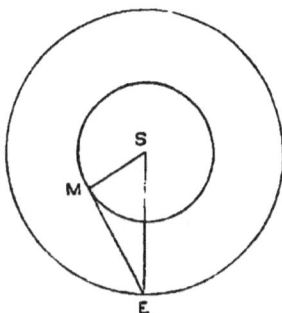

Fig. 93.

208. The aphelion distance of Mercury from the sun is about 42,600,000 miles; its perihelion distance, 28,100,000 miles; and its mean distance, therefore, nearly 35,400,000 miles.

a. **The Distance Calculated.**—The distance of an inferior planet can be determined by its extreme elongation; for (Fig. 93) S M is equal to S E multiplied by the sine of the angle S E M. Now, if the angle is 28° 15', its sine (or ratio of S M to S E) will be .473, and hence, taking the perihelion distance of the earth, 90,000,000 × .473 = 42,600,000, which is the aphelion distance of Mercury.

b. The mean distance of any planet from the sun can be calculated by Kepler's third law, when we know the sidereal period. In the case of Mercury, this is very nearly 88 days; and hence, as *the squares of the periodic times are in proportion to the cubes of the mean distances,* $(365\frac{1}{4})^2 : (88)^2 :: (91,500,000)^3$: the cube of the mean distance of Mercury, which, if the proportion be worked out and the cube root extracted, will give 35,400,000 (nearly); and this is the true value of this element. [If the pupil is sufficiently advanced, it will be well for the teacher to show how this calculation may be facilitated by employing a table of common logarithms.]

c. The difference between the aphelion and perihelion distances of Mercury, it will be seen, is 14,500,000; hence its eccentricity is 7,250,-000 miles, which is nearly .205 of its mean distance, or about 12 times as great as that of the earth.

QUESTIONS.—208. What are the aphelion, perihelion, and mean distances of Mercury? *a.* How calculated? *b.* How determined by Kepler's third law? *c.* Eccentricity—how found?

209. The apparent diameter of Mercury, when greatest, is about 13 seconds, and when least, about $4\frac{1}{2}$ seconds; this difference being caused by the variations in its distance from the earth.

a. When in inferior conjunction it is, of course, at the point nearest to the earth, and when in superior conjunction, at its farthest point from the earth. When Mercury is in superior conjunction and each planet is at its aphelion, they are the farthest possible from each other; that is, $42,600,000 + 93,000,000 = 135,600,000$ miles. When Mercury is in inferior conjunction and at its aphelion, while the earth is in its perihelion, they are nearest to each other; that is, $90,000,000 - 42,600,000 = 47,400,000$ miles.

210. The real diameter of Mercury is about 3,000 miles (more exactly, 2,962 miles).

a. This is deduced from its distance and apparent diameter by the method explained in Art. 147, *a.* Suppose the apparent diameter is ascertained to be $12\frac{1}{4}''$, while its distance from the earth is 50,000,000 miles. Then, the sine of $6\frac{1}{8}''$, which is .00002962, being multiplied by 50,000,000 gives 1481, the semi-diameter.

b. A clearer idea may be formed of the apparent diameter of a planet by comparing it with that of the moon, which subtends an angle of more than $1,800''$; consequently, Mercury, when it appears as a thin crescent near inferior conjunction, subtends an angle equal to only about $\frac{1}{140}$ or $\frac{1}{150}$ part of the moon's disc; while when it recedes to its greatest distance, it is about $\frac{1}{400}$ part.

c. The oblateness of the planet has been generally considered very small; but one astronomer (Dawes), in 1848, gave it at $\frac{1}{38}$, which would

Fig. 94.

M ... E

COMPARATIVE VOLUMES OF MERCURY AND THE EARTH.

be nearly ten times as great as that of the earth.

d. **Volume, Mass, and Density.**—The volume of Mercury is about .052 of the earth's; and the mass has been estimated at about .063; hence, its density must be $.063 \div .052 = 1.12$, the earth's being 1;

and since this is 5.67 of water, 5.67 × 1.12 = 6.35, must be the density as compared with water. To find the mass of Mercury is so difficult a problem, that these figures can not altogether be relied on. There is but little doubt, however, that its density exceeds that of the earth by an eighth to a fifth. The famous French mathematician Le Verrier estimates it at more than twice that of the earth.

e. **Superficial Gravity at Mercury.**—Since gravity varies directly as the quantity of matter, and inversely as the square of the distance from the centre, at the surface of Mercury it must be nearly ($\frac{1288}{285}$)² × .063, or $\frac{81}{4}$ × .063 = .448 (nearly). Hence, a body that weighs a pound at the surface of the earth would weigh less than half a pound at Mercury. If we should be transported to this planet, we should appear to have more than twice as much muscular power, since the resistance to our efforts would be diminished more than one-half.

211. Mercury is supposed to perform a diurnal rotation in about 24 hours (24ʰ 5ᵐ 28ˢ).

a. This was discovered by the celebrated German astronomer, Schroeter, at the end of the last century, by examining daily the appearance of the *cusps*, or extremities of the crescent form of the planet, which instead of being pointed are sometimes obtuse, owing to irregularities on the surface of the planet, and during one rotation undergo certain changes in form, which enabled the astronomer to discover the period. He also thought that he had discovered certain dark bands across the disc, and was enabled to measure the height of some of the mountains, which, according to his computations, are very high, some of them more than ten miles,—an enormous altitude for so small a body. These discoveries have not, however, been confirmed by the observations of other astronomers. Sir John Herschel, with all his advantages for telescopic observation, and his great experience and skill as an observer, states that "all that can be certainly affirmed of Mercury is, that it is globular in form and exhibits phases ; and that it is too small and too much lost in the constant and close effulgence of the sun to allow the further discovery of its physical condition."

b. Mercury is a very difficult object to see in consequence of its nearness to the sun ; for it is generally entirely involved in the twilight or obscured by the mists that float near the horizon. Copernicus,

it is said, regretted at his death that he had never been able to obtain
a view of it. In lower latitudes, where the diurnal circles are more
nearly vertical, the twilight shorter, and the atmosphere less clouded,
it is more easily seen. The fact that it was so well known to the
ancients as a planetary body proves to us that they were very careful
and diligent observers. When viewed through a telescope, it looks
intensely brilliant, on account of its proximity to the sun ; and this
excessive brilliancy serves to prevent the clear observation of any spots
on the disc which would afford positive indications of its axial rotation.
It has, nevertheless, been very diligently observed by astronomers.
Arago states that at the observatory of Paris alone, more than two hundred complete observations of it were made from 1836 to 1842.

212. Mercury performs its sidereal revolution in about 88
days (87^d 23^h 16^m) ; its synodic period is about 116 days.

a. The sidereal period may be ascertained by observing when the
planet is at either node, and noticing the interval of time that elapses
before it returns to the same node. The synodic period can be calculated on the principle already explained. Thus, $365\frac{1}{4}$ days $\div$
87.968 days $= 4.152$, which is the number of sidereal revolutions in a
year ; hence there must be 3.152 synodic revolutions, or one less than
the sidereal ; and $365\frac{1}{4}$ days $\div$ $3.152 = 116$ days (nearly).

Or the synodic period may be found by observation, and the sidereal
period deduced from it on the same principle. ($365\frac{1}{4} \div 116 = 3.152$ $\therefore$
$365\frac{1}{4} \div 4.152 = 87.968$.)

213. The apparent diameter of the sun as seen at Mercury varies considerably, owing to the great difference in its
distance at aphelion and perihelion. When in the former
position it is nearly 69′ ; in the latter, 104′ ; being when
least more than twice, and when greatest more than $3\frac{1}{2}$
times that of the sun as seen from the earth.

a. Light and Heat at Mercury.—The light and heat received
from the sun when Mercury is in perihelion must be greater than in
aphelion, in the proportion of 2.27 to 1 ; that is, in the former, about
$2\frac{1}{4}$ times as great as in the latter ; for the light and heat vary in

QUESTIONS.—212. What is the sidereal period of Mercury ? Its synodic period ? *a.*
How may one be deduced from the other ? 213. Apparent diameter of the sun at Mercury ? *a.* Its light and heat ?

proportion to the area of the sun's disc, and this is as the square of the apparent diameters; or as $(104)^2$ to $(69)^2$, or as 2.27 to 1. On the same principle, the average amount of light and heat received by Mercury, is to that received by the earth as the square of the mean apparent diameter of the sun at that planet $(83')^2$ is to that at the earth $(32')^2$; that is, as 6889 is to 1024 or as $6\frac{3}{4}$ (nearly) to 1. In other words, the light and heat at Mercury are nearly $6\frac{3}{4}$ times as great as at the earth. This, of course, may be very much modified by other circumstances.

b. **Seasons at Mercury.**—Since the light and heat are more than $2\frac{1}{4}$ times as great in perihelion as in aphelion, there must be a succession of seasons on the planet depending entirely on the eccentricity of its orbit, summer occurring when the planet is in perihelion, and winter when it is in aphelion. If the axis of the planet is inclined to its orbit, another succession of seasons must occur, which, if they coincide with the former, must, in one hemisphere, augment the intensity of the heat, and in the other, increase that of the cold; while, if they do not coincide, the one cause must tend to counteract the effects of the other, and thus diminish the great extremes of heat and cold.

214. Transits of Mercury.—When the latitude of Mercury or Venus, at the time of inferior conjunction, is less than the semi-diameter of the sun, a *transit* must occur, the planet appearing on the sun's disc like a small, round, and intensely black spot, and moving across it from east to west.

a. It appears to move across the disc from east to west for the same reason that the solar spots appear to move in that direction (Art. 152, *b*). The planet's velocity being faster than the earth's, the planet passes the earth actually from west to east, but in a direction opposite to the diurnal motion of that part of the earth on which the observer stands; hence the apparent motion is *westward*, since east is in the direction of the rising sun, and this must be the point toward which any place on the earth turns.

b. **Transit Limits.**—The limits in latitude within which a transit can occur correspond to the semi-diameter of the sun; and as Mercury's orbit is inclined to the ecliptic at an angle of 7°, it can easily be computed that the planet, at inferior conjunction, must be within about 2° of longitude from the node for a transit to occur.

c. **Times of the Occurrence of Transits.**—The longitudes of Mercury's nodes are 46° (☊) and 226° (☋); hence they are in Taurus and Scorpio, and the earth arrives at the line of nodes in May and November of each year. Transits must, consequently, occur in these months; and this will continue to be the case for a long time, because the nodes change their position on the ecliptic only about 13' in a century. Because 7 sidereal periods of the earth are very nearly equal to 29 (29.064) of those of Mercury; and 13 of the former are nearly equal to 54 (53.98) of the latter, transits must, as a general thing, occur at intervals of 7 or 13 years, at the same node. Owing, however, to the great inclination of the orbit, and the consequent small transit limit, these periods can not be relied on; and it requires a period of 217 years to bring round the transits exactly in the same order. The last transit occurred on the 12th of November, 1861; the next will occur on the 5th of November, 1868.

II. VENUS. ♀

215. VENUS is in appearance the most brilliant and beautiful of all the planets, and is remarkable for its close resemblance to the earth, both in size and mass.

a. **Name and Sign.**—Venus, in the pagan mythology and religion, was the goddess of beauty, and hence the name was appropriately applied to the most beautiful of the planets. The sign is supposed to represent a mirror having a handle at the bottom. By the ancients, this planet, when an evening star, was called *Hesperus* or *Vesper*, the former being a Greek word, and the latter a Latin word, each meaning *the evening*. When a morning star, it was called *Phosphorus* or *Lucifer*, the former, in Greek, signifying *that which* *brings the light;* and the latter meaning the same thing in Latin. These were at first supposed to be different bodies.

216. The PHASES of Venus exactly resemble those of the

moon, and when viewed with a telescope are very interesting and beautiful, clearly proving that this planet revolves within the earth's orbit.

217. When the planet is in superior conjunction its full disc is visible, which gradually diminishes until, at the time of greatest elongation, only half of the disc is seen; after which the planet still continues to wane until, when near inferior conjunction, it assumes the form of a slender crescent.

218. When the planet is full, its apparent diameter is least, since it is then farthest from the earth; but near inferior conjunction, its apparent diameter is *greater* than at any other time, except when it is seen during a transit.

The accompanying diagram (Fig. 95) shows the phases of Venus during one synodic period. The difference in size when the planet is full and when it has the crescent form, will be obvious. This difference, however,

Fig. 95.

PHASES OF VENUS.

is greater than here presented; the apparent diameter in inferior conjunction being more than six times as great as it is when in superior conjunction.

219. The extreme elongation of Venus never exceeds
47¾°, its average being about 46°.

a. This small amount of variation indicates that the orbit has very
little eccentricity.

b. Venus is most brilliant when the elongation is about 40°, and
when its apparent diameter is about 40″, and about ¼ of the entire
disc is visible, this portion giving more light than phases of greater
extent, because the latter are presented at so much greater distances.
During every eight years, when the planet has its greatest north latitude and is 40° from the sun, its brilliancy is so dazzling that it is
visible in full daylight, and casts a sensible shadow in the evening.
This was the case in Feb. 1862, and had been often observed previously.

220. The aphelion distance of Venus from the sun is
about 66,600,000; the perihelion distance, 65,700,000; its
mean distance being about 66,150,000.

a. The mean distance can be calculated by Kepler's third law ; and
the aphelion and perihelion distances by the greatest and least extreme
elongations, as in Art. 208.

☞ [The pupil should be required to make the calculations in each
case. The ratios or sines of the angles can be found by consulting any
ordinary table of natural sines. The term *natural* is employed to distinguish these ratios from the *logarithmic* expression of them.]

b. The absolute eccentricity of Venus, it will be seen, is only about
450,000 miles, or less than .0069 of its mean distance. So nearly does
the orbit resemble a circle.

221. The apparent diameter of Venus, when greatest, is
66½″; and when least, somewhat less than 10″. At its
mean distance (91 millions of miles) from the earth it is
about 17″; and therefore its real diameter is 7510 miles.
(Art. 210.)   The compression or oblateness of the planet
is exceedingly small.

a. **Volume, Mass, and Density.**—The volume of Venus is about
.85 of the earth's ; while its mass is .89 ; hence its density must be

QUESTIONS.—219. Extreme elongation? *a.* Why but slightly variable? *b.* When is
Venus most brilliant? 220. Distances of Venus? *a.* How calculated? *b.* Eccentricity? 221. Apparent and real diameter? *a.* Volume, mass, and density?

.89 + .85 = 1.047 (nearly), or very lfttle greater than that of the earth.

b. **Superficial Gravity.**—This is found as in Art. 210, *e,* $(\frac{3}{2}\frac{0}{7}\frac{9}{2})^2 \times$ .89 = .98 ; that is, a pound at the earth's surface would weigh only .02 less at the surface of Venus.

222. The diurnal rotation of Venus is performed in 23^h 21^m 19^s.

a. This is the period as determined by Schroeter in 1789, by discovering a luminous point in the dark hemisphere a little beyond the southern horn of the planet, indicating the existence there of a high mountain. By watching its periodic changes, he was enabled to deduce the time of the rotation ; which agreed very nearly with the result attained in 1667, by Cassini, and was subsequently confirmed by the observations of other astronomers.

b. **Mountains in Venus.**—According to Schroeter, there exist mountains of immense height on the surface of Venus ; the elevation of the loftiest being equal to $\frac{1}{140}$ of the planet's radius, which would be about 27 miles, or five times the altitude of the loftiest terrestrial peak.

Fig. 96.

TELESCOPIC VIEWS OF VENUS.

While other astronomers have detected the existence of mountains on the planet, their observations do not confirm the statements of Schroeter as to their height. Very great irregularities on its surface are, however, indicated by the jagged character of the terminator, by the shading of its edge, and by the blunt or broken extremities of its cusps. The shading is supposed to be caused by the long shadows cast by the mountains as the sun shines obliquely upon them. [See Fig. 96.] The telescopic observation of Venus is attended with great difficulty on account of the intense brilliancy of its light, which dazzles the eye and augments all the imperfections of the instrument.

223. There are undoubted indications that Venus is surrounded by an atmosphere of considerable height and density.

a. This is shown by several circumstances: 1. When the planet was seen as a narrow crescent, Schroeter remarked a faint light projecting beyond the proper termination of one of the horns into the dark part of the planet. This has been seen by other observers, and is supposed to be due to the existence of an atmosphere refracting the light of the sun. 2. By observing the concave edge of the crescent, it is found that the enlightened part does not terminate suddenly, but that there is a gradual fading away of the light into the dark portion of the planet's surface, this being obviously occasioned by a reflection of the sun's rays, producing the phenomenon of twilight. 3. During the transits of 1761 and 1769, the planet was observed, by several astronomers, to be surrounded by a faint ring of light, caused, as it has been supposed, by the sun's rays passing through the planet's atmosphere. 4. *Clouds* have been observed floating over the disc of the planet, and screening by their greater brilliancy its darker surface, which is occasionally seen between them. Such being the case, there must be *water* as well as *air* on the surface of Venus.

224. The inclination of the axis of Venus has not been positively ascertained; but it is supposed to be very great, according to several astronomers, about 75°.

a. This can be positively discovered only by observing spots on the disc, and noticing the direction of their apparent motion. This, in the case of Venus, is exceedingly difficult, owing to the intense brilliancy of its light and the density of its atmosphere. Cassini, as early as 1666-7, saw one bright and several dusky spots, and others have been observed since that time by different astronomers, among whom, De Vico, at Rome, in 1839-41, appears to have attained the most reliable results.

225. If the axis of Venus has an inclination of 75°, its tropics must be 75° from its equator, and its polar circles 75° from the poles. Hence it can have only a torrid zone, which must be 150° wide, and frigid zones extending 75° from the poles.

226. SEASONS OF VENUS.—As the sun must arrive at the equator and depart to its greatest distance from it twice during each sidereal period, there must be two summers and two winters at this part of the planet, and a summer and winter at each of the poles, which must suffer a transition from the burning heat of a vertical sun and constant day, to the intense cold of perpetual night, each lasting more than 112 days.

Fig. 97.

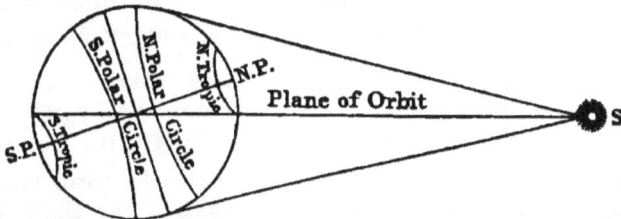

Fig. 97 exibits the relative positions of the tropics and polar circles, the former being the nearest to the poles. The sun is represented as in the northern solstice; and it will be seen that all places situated more than 15° north of the equator have constant day, and those more than 15° south of it, constant night. Hence there must be winter at the equator and within the south polar circle, and summer within the north polar circle. In one-fourth of the year, when the sun will have arrived at the equator, there will be equal day and night all over the planet, summer at the equator, autumn within the north polar circle, and spring within the south polar circle.

Fig. 98 exhibits the planet at each of the equinoxes and solstices: To an inhabitant of the northern hemisphere of Venus, at A the sun is in the vernal equinox; B, the summer solstice; C, the autumnal equinox; and D, the

winter solstice. To an inhabitant of the southern hemisphere, these
would, of course, be reversed.

Fig. 98.

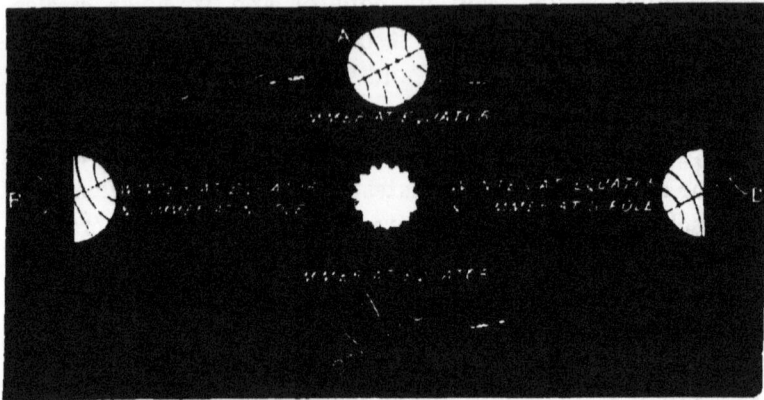

SEASONS OF VENUS.

227. Venus performs its sidereal revolution in about $224\frac{2}{3}$
days (224^d 16^h 49^m) ; its synodic period is about $584\frac{1}{2}$ days.

a. For its sidereal period is 224.7 days, and $365.25 \div 224.7 = 1\frac{5}{8}$;
hence $1\frac{5}{8} - 1 = \frac{5}{8}$, is the part of a synodic revolution performed in
$365\frac{1}{4}$ days ; and $365\frac{1}{4} \div \frac{5}{8} = 584\frac{1}{2}$ days. [See Art. 212, *a.*]

b. **Division of the Synodic Period.**—Since the mean value of
the extreme elongation of Venus is 46°, which is the angle M E S
(Fig. 93), the angle at the sun M S E must be 90° — 46° = 44°. Therefore the planet, in passing from M to the other point of extreme
elongation, has to gain on the earth 88° ; and the time required
for this must bear the same ratio to $584\frac{1}{2}$ days as this angle bears to
360°. Consequently, the synodic period is divided into the following
intervals : $\frac{88}{360} \times 584\frac{1}{2}$ days = 142.9 days, the interval between the time
of greatest elongation before and after inferior conjunction, and $\frac{272}{360}$
$\times 584\frac{1}{2}$ days = 441.1 days, the interval between the elongation before
and after superior conjunction. One-half of each of these intervals
will give the time from the greatest elongation to inferior conjunction,
and that from greatest elongation to superior conjunction, respectively.

QUESTIONS.—227. Sidereal period ? Synodic period ? *a.* How calculated ? *b.* Interval of time between the elongations and conjunctions ?

c. **Morning and Evening Star.**—Since Venus must remain on each side of the sun during 292 days, or one-half the synodic period, it must be a morning and evening star alternately for that time

228. The distance of Venus from the sun being a little more than $\frac{7}{10}$ the distance of the earth, the apparent diameter of the sun must be $1\frac{3}{7}$ as great, or about 45′. Hence the solar light and heat must be more than twice as great.

a. For these vary inversely as the squares of the distances; and as the ratio of the distances is .723, $(.723)^2$, or about .52 will be the ratio of the light and heat of the earth to those of Venus. This, of course, may be very much modified by the influence of its atmosphere.

229. TRANSITS OF VENUS.—The orbit of Venus is inclined to the ecliptic at an angle of $3\frac{1}{2}°$ (3° 23′ 29″); and therefore a transit can take place only when the planet is at or near one of its nodes. The transits of Venus are of great interest and importance, because they afford a means of determining the parallax of the sun, and consequently its distance from the earth.

a. **History of Transit Observations.**—These transits are of rare occurrence. Kepler, in 1629, predicted that a transit would occur in 1631, and that no other would occur until 1761. His prediction, however, was not confirmed by observation, for the transit of 1631 occurred during the night; and in 1639 the *first observation of a transit* was made by a young English astronomer, named Horrox. No other occurred till 1761 and 1769; and these were observed with great care in different parts of the world, in order to apply the method of finding the solar parallax, first suggested by Gregory, a celebrated mathematician, in 1663. King George III., in 1769, despatched, at his own expense, an expedition to Otaheite, under the command of the celebrated navigator, Captain Cook, in order that observations of the transit might be taken in this distant spot. Other nations sent in different directions similar expeditions. The average result of these different observations assigned as the true solar parallax 8.5776″, which has, within a very few years, been found to be somewhat too small. The next transit will occur in 1874.

b. **Times of the Occurrence of Transits.**—The longitudes of the nodes of Venus are about 75° (Ω), and 255° ($\mho$): hence, they are in the 15th degree of Gemini and Sagittarius ; and as, therefore, the earth arrives at the line of nodes early in June and December, the transits must occur in these months ; and will continue to do so for a long time, since the longitude of the nodes diminishes, according to Le Verrier, at the rate of less than 18' in a century.

c. Because 8 sidereal periods of the earth are very nearly equal to 13 of Venus, transits often occur at intervals of eight years ; when, however, two transits have occurred at this interval, another can not be expected before $105\frac{1}{2}$ years. Thus, the next transit will happen in December, 1874 ; and another in $1874 + 8 = 1882$ (December 6th). The next will not occur till June 7th, 2004, $121\frac{1}{2}$ years afterwards. The transits are thus repeated at intervals of 8, $105\frac{1}{2}$, 8, $121\frac{1}{2}$, 8 years, etc. : acccording as they occur at one or the other node.

d. **How to Find the Sun's Parallax by the Transits of Venus.**— By the greatest elongation, the distance of an inferior planet from the sun can be found, provided we know the earth's distance. We, in fact, only find by this method the *ratio* of one distance to the other ; and this is known, therefore, independently of the solar parallax.

Fig. 99.

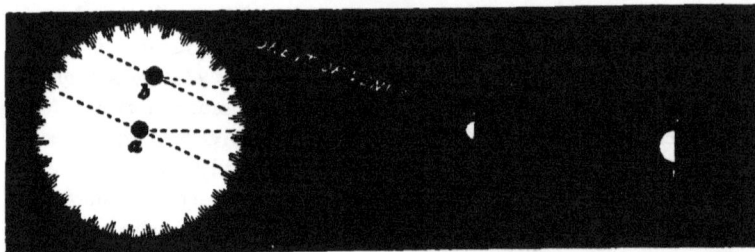

TRANSIT OF VENUS.

In Fig. 99, let A and B be the positions of two observers stationed at opposite parts of the earth, and V, the place of Venus at the time of a transit. The observer at A will see the planet projected on the sun at *a*, and the observer at B, at *b*; and if each observer notice exactly the time required by the planet to cross the disc, he can, since its hourly motion is known, easily calculate the length of the chord which it appears to describe at each place ; and a comparison of the length of these chords will give the distance between a and b in seconds of space. Now, if Venus were exactly half-way between the earth and sun the distance $a\,b$ as seen from the earth would be exactly the same as A B seen from the sun ; and therefore would be twice the parallax. But the distance of Venus from the sun is known to be .723 of the earth's distance; and consequently A V, its distance from the earth, must be 1 —.723 = .277 ; so that the ratio of a V to A V is $\frac{1.000}{.277}$ = 2.6 (nearly). But $a\,b$ bears the same ratio to A B as a V does to A V : hence, the distance $a\,b$ must be 2.6 × 2 or 5.2 times the solar parallax. Suppose, therefore, this distance should be found by observation and calculation to be 46½″ ; then 46½″ ÷ 5.2 = 8.94″ would be the true parallax. The advantage afforded by this method is, that because the distance of Venus from the earth is so much less than from the sun, $a\,b$ is enlarged in the same proportion, and thus rendered more susceptible of exact measurement. Thus, it will be obvious that whatever error arises in determining $a\,b$, affects the parallax less than one-fifth. Practically, it is impossible that the observers should be as far apart as A B ; but whatever their distance from each other, it can be easily reduced to the conditions represented in the diagram.

APPARENT MOTIONS OF THE INFERIOR PLANETS.

230. The APPARENT MOTIONS of Mercury and Venus are sometimes from west to east, and sometimes from east to west. The former are said to be *direct ;* the latter, *retrograde.* At certain intermediate points, the planet appears to remain for a short time in the same point of the heavens, and is then said to be *stationary.*

a. The student must clearly understand that these motions and stationary points have reference to the *stars,* not to the *sun ;* the apparent place of the latter is constantly changing on account of the motion of the earth. These phenomena are illustrated in the annexed diagram.

Let S (Fig. 100), be the place of the sun, the inner circle the orbit of an inferior planet ; the outer circle that of the earth. Let also *a, b, c, d,* etc., be the positions of the planet at unequal intervals of time between the points

---

of extreme elongation, a and g; and A, B, C, D, etc., the places of the
earth at the same time; while 1, 2, 3, 4, etc., represent the apparent places
of the planet, as seen in the sphere of the heavens.   In passing from g,
the western point of extreme elongation, through o, the place of superior

Fig. 100.

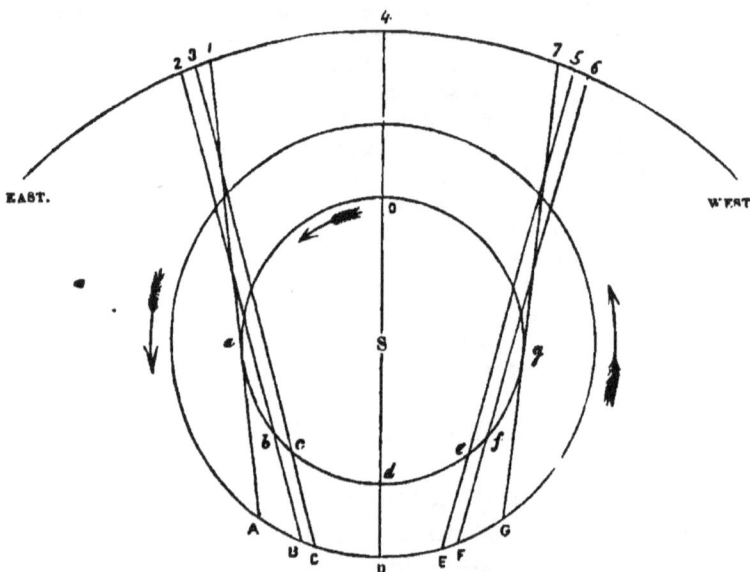

APPARENT MOTIONS OF VENUS AND MERCURY.

conjunction, to a, the eastern point of extreme elongation, the planet evidently must appear to move toward the east; and when it arrives at a, the
earth being at A, it still continues to be direct for a short time; for while
going from a to b its motion is so oblique that the earth passes it, so that
when the latter arrives at B, the planet appears to have moved from 1 to 2.
Its elongation is not, however, increased since the sun itself has moved farther to the east.   While the planet is going from b to c, and the earth from
B to C, the former does not appear to change its position at all; for the
lines B b 2 and C c 3 are parallel, and consequently indicate no change of place
among the stars, and 2 is to be considered, therefore, as identical with 3.
The reason of the planet's appearing stationary, it will be seen, is that the
obliquity of its motion exactly counterbalances the difference between its
actual velocity and that of the earth; b is, therefore, to be considered the
stationary point. At d, the planet is in inferior conjunction, having overtaken
the earth, and is seen at 4, to the west of its previous position.   In passing

from *d* to *g* the same phenomena are presented in the reverse order; at *e* it becomes stationary, remaining so till it reaches *f*, where it ceases to be retrograde, appearing to move while going from *f* to *g*, from 6 to 7. In going from *c* to *e*, the two stationary points, it has evidently changed its direction among the stars, not by the actual distance 3, 5, but by the angle contained by the lines 3 *c* C and 5 *e* E when produced until they meet in some point below C D E. This angle, or the arc by which it is subtended, it is obvious, is quite small; it is called the *arc of retrogradation*. From the above explanation the following statements will be understood.

231. An inferior planet appears stationary at two points of its synodic revolution, between the extreme elongations and inferior conjunction. Its motion is retrograde in passing through inferior conjunction from one stationary point to the other; and direct in passing through superior conjunction, between the same two points.

a. The stationary points of Mercury vary from 15° to 20° of elongation from the sun ; those of Venus generally occur when its elongation is about 29°. (See Fig. 95).

b. The time during which Mercury retrogrades is about 22 days; Venus, 42 days. The mean arc of retrogradation of the former is about 12½° ; of the latter, 16°. The stations of both are, of course, but of very short duration.

QUESTIONS.—231. When does an inferior planet appear stationary? When is its motion direct? When, retrograde? *a.* Where are the stations of Mercury and Venus? *b.* During how many days does each retrograde? How long are the arcs of retrogradation?

CHAPTER XIII.

I. MARS. ♂

232. MARS, the fourth planet from the sun (the most distant of the terrestrial planets), is remarkable for its small size and the red color with which it shines among the stars.

a. **Name and Sign.**—This redness of its appearance makes it easily distinguished among the other heavenly bodies, and doubtless gave rise to its name; Mars, in the heathen mythology, being the god of war. Its sign is a shield and spear.

233. PHASES.—When Mars is in opposition or conjunction, its disc is full; when between these points, it is gibbous. More than half of its disc is, therefore, always visible.

a. The reason of this will be apparent after a little consideration. In opposition and conjunction, the hemisphere presented to the earth exactly coincides with the illuminated disc; hence, the planet appears full. The amount of diminution of the full disc is obviously equal to the angle formed at the centre of the planet by lines drawn to the earth and sun. (See Fig. 75.) As, in the case of a superior planet, this angle is always less than a right angle, the amount of dimunition must be less than one-half of the disc; for a right angle would include one-quarter of the whole surface, which is, of course, one-half of the disc. Therefore, the planet can present no other than the full or gibbous phase.

234. The APPARENT MOTIONS of a superior planet, like

QUESTIONS.—232. For what is Mars remarkable? *a.* Name and sign? 233. What phases does it exhibit? *a.* How is this explained? 234. What apparent motions have the superior planets?

those of an inferior planet, are either direct or retrograde; and as its motion changes from one to the other, it appears for a short time to be stationary.

235. The motion appears to be retrograde for a short distance before and after opposition, and direct in the other part of its orbit. The retrogradation of the planet is caused by the greater velocity of the earth; so that as the latter body moves toward the east, it passes the other, and thus makes it appear to move toward the west. When the motion of the earth is sufficiently oblique to counteract the excess of its velocity, the two bodies move on together, and the planet appears to be stationary.

a. The *arc of retrogradation* in the case of the superior planets is very small. The following is the mean value of each: Mars, 15°; Jupiter, 10°; Saturn, 6¾°; Uranus, 3¾°; Neptune, about 2°.

b. Mars retrogrades from 60 to 80 days, according as it is in perihelion or aphelion; Jupiter continues to retrograde during about 4 months; Saturn, about 4½ months; Uranus, about 5 months; Neptune about 6 months.

c. Mars becomes stationary when its elongation is about 140°; Jupiter, 115°; Saturn, 110°; Uranus, 103°; Neptune, 97½°.

236. The aphelion distance of Mars is about 152,300,000 miles; its perihelion distance, 126,300,000; hence, its mean distance is about 139,300,000.

a. Since the periodic time of Mars is $1^y 322^d$, we have, by Kepler's third law, $(365\frac{1}{4})^2:(687\frac{1}{4})^2::(91,500,000)^3$: the cube of the distance of Mars, which, by working out the proportion, will give very nearly the mean distance above stated.

b. The distance can also be found approximately by observing the phase of Mars when it varies most from the full. The following is the method:—

Let S (Fig. 101), represent the sun, E, the earth, and M, Mars, much

Fig. 101

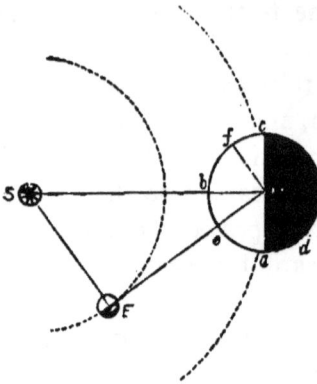

enlarged for convenience of illustration. The earth is obviously at the
point of greatest elongation as seen
from Mars; and hence the angle
S M E is the largest possible. But this
is equal to a M d which measures the
deviation of the disc from the full;
for the arc $b\,a = e\,d$; and taking $e\,a$
from each, there remain $b\,e = a\,d$. Suppose this angle is measured and found
to be 41° 4'. Then the sine of this angle
being about .657, this must be the
ratio of the distance of the earth to
that of Mars; and $91,500,000 \div .657 =$
$139,300,000$ (nearly). This method is
applicable only to Mars; since the
other planets are so far distant that the angle of elongation corresponding
to S M E, is very small; and their deviation from the circular outline not
large enough to be susceptible of exact measurement.

c. A more general method, based upon the daily arc of retrogradation of the superior planets, is interesting as having been employed
by the old astronomers, and more particularly by Kepler in those
investigations which resulted in the discovery of his third law. The
following is a brief statement of this method :—

Let S (Fig. 102), represent the sun, E e the daily arc of movement of the
earth, and M m, that of Mars in opposition. In going from M to m, the
planet obviously changes its direction backward among the stars by the
angle $d\,e\,o = e\,o$ E. Now, in the triangle $o\,e$ S, the angle o is given, also the

Fig. 102.

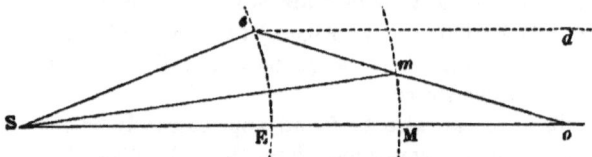

angle E S e, since it is the amount of angular movement of the earth
for the time; hence, the third angle $o\,e$ S becomes known; and the ratio of
o S to e S, since this is equal to the ratio of the sines of $o\,e$ S and $e\,o$ S. In
the triangle $o\,m$ S we have, in the same manner, the angles o and m S M,
the latter being the angular movement of the planet, so that the ratio of

---

QUESTION.—*c.* How may the distance be found by the arc of retrogradation?

o S to *m* S becomes known; whence, combining the two results, we obtain the ratio of *m* S to *e* S, or that of the two distances; on the supposition, however, that the orbits are circular; but when the process is repeated in every variety of situation at which the opposition may occur, the average of the results will give a tolerably accurate determination.

237. The ECCENTRICITY of the orbit of Mars is about 13 millions of miles, or .093 of its mean distance. It is, therefore, nearly 5½ times as great as that of the earth.

a. **Parallax of Mars and of the Sun.**—Owing to the great eccentricity of the orbit of Mars, it sometimes, when in opposition, approaches very near to the earth ; for if it is in perihelion while the earth is in aphelion, the distance is 126,300,000—93,000,000 = 33,300,000. Advantage was taken of this in 1862 to determine its parallax. It was arranged that several observers should station themselves at places in different hemispheres and record the zenith distance of the planet when on the meridian, as well as its distance from certain stars, so as to ascertain the amount of displacement in its position occasioned by the separation of the observers. These observations were made at Greenwich, Pulkova, Washington, Cape of Good Hope, Williamstown in Australia, and Santiago.

An effort was also made to determine the parallax at a single observatory, by observing the displacement in the apparent position of the body, occasioned by the rotation of the earth. For as the observer is not situated at the centre of motion, the planet can only appear precisely in its true place, as to right ascension, when it is on the meridian ; and frequent observations made before and after culmination must show an average displacement, from which its parallax may be calculated, and consequently its distance from the earth. From this may be deduced the earth's distance from the sun, and of course the solar parallax. The results of the different observations very nearly agreed, all showing the parallax as determined in 1769 to be too small. The average result is that now generally accepted (8.94″). Le Verrier had previously assigned very nearly the same amount by calculations based upon the disturbances of the planets, which showed that the parallax needed correction in order to bring the observed perturbations into harmony with those theoretically computed.

238. The INCLINATION OF THE ORBIT of Mars to the plane of the ecliptic is only about two degrees (1° 51′).

239. Its SIDEREAL PERIOD is nearly 687 days; its synodic period, 780 days.

a. For $687^d - 365\frac{1}{4}^d = 321\frac{3}{4}^d$; hence, $687^d \div 321\frac{3}{4}^d = 2.135$, is the number of revolutions of the earth during the synodic period; and $365\frac{1}{4}^d \times 2.135 = 780^d$ (nearly). In Art. 62, the same method is applied *fractionally ;* thus, $365\frac{1}{4}^d = .532$ of 687^d; hence, the earth gains in one revolution $1 - .532 = .468$ of a revolution upon the planet ; but she has to gain an entire revolution ; and $1 \div .468 = 2.135$ revolutions (nearly).

240. The apparent diameter of Mars varies between 4″ in conjunction and 30″ in opposition. Its real diameter is about 4,300 miles. Its oblateness is about $\frac{1}{50}$ of its diameter, or 86 miles, and is consequently very nearly six times as great as the earth's.

241. It performs its daily rotation in about 24$\frac{1}{2}$ hours (24^h 37^m 42^s), upon an axis inclined toward its orbit 28° 42′; hence its obliquity is nearly the same as the earth's, and its variety of seasons also the same, except that they are nearly twice as long.

a. Seasons of Mars.—The year of Mars contains 668^J 16^h of its own time, since its days are longer than those of the earth. Owing to the great eccentricity of its orbit, summer in the northern hemisphere is only about $\frac{2}{3}$ as long as in the southern ; but in consequence of its greater proximity to the sun, the light and heat are much greater (in the ratio of 145 to 100). Thus there is a complete compensation in the seasons of both hemispheres. Constant day at the north pole of Mars lasts during 297 of its days ; at the south pole, during 372 days. Hence, *constant night* at the north pole is 75 days longer than at the south pole.

242. The TELESCOPIC APPEARANCES of Mars are very

interesting, exhibiting what seem to be the outlines of continents and seas, the former appearing of a ruddy or orange color, and the latter of a dusky greenish or bluish tint.

243. Brilliant white spots are also seen alternately at the poles, produced, as it is conjectured, by accumulations of ice and snow during the long winters, particularly as they are seen to disappear as summer advances upon the poles.

244. Evidences are also presented of an atmosphere, probably of a density about equal to that of the earth.

Fig. 103.

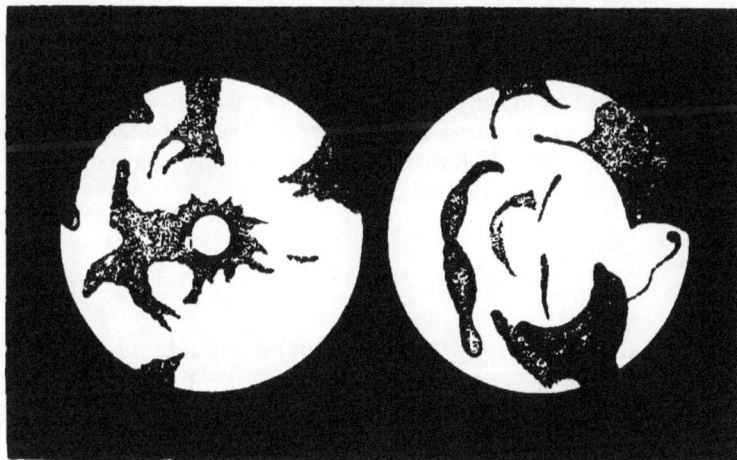

NORTHERN AND SOUTHERN HEMISPHERES OF MARS.—*Mädler.*

a. With the exception of the moon, no body has been submitted to such a careful telescopic scrutiny as Mars. The utmost assiduity has been exercised particularly by Messrs. Beer and Mädler in these researches, which were commenced by them in 1830, and continued at every opportunity for twelve years. A large collection of drawings of the various hemispheres presented by the planet was made by them, showing the positions and outlines of the spots seen on the disc, and clearly establishing their connection with the planet's

surface, and their general permanency. Considerable variety is, however, exhibited in the forms of these spots, owing to the great diversity
in the hemispheres presented.

The annexed cut does not represent any actual telescopic views of the
planet, since we are never so situated as to be able to see the *whole* of either
the northern or southern hemisphere at any one time. It exhibits a combination of a large number of telescopic appearances, the various dusky spots
being placed together so as to show the forms of the *different bodies of water*
and their relation to the continents; the latter being indicated by the white
spaces. These, through the telescope, appear of a ruddy color, and give
this general tint to the planet. On the earth, the continents are islands,
being encompassed by the water; on Mars, it will be perceived, the bodies
of water are lakes or seas, being entirely encompassed by the land.

b. No entirely satisfactory cause has been assigned for the ruddy
color of this planet. It is thought by Sir John Herschel to be due to
" an ochrey tinge in the general soil, like what the red sandstone dis-
tricts on the earth may possibly offer to the inhabitants of Mars,
only more decided." Viewed through a telescope, the redness of its
hue is very considerably diminished.

II. JUPITER. ♃

245. JUPITER, the first of the major planets, is remarkable
for its great size, it being the largest of all the planets.
It is also distinguished for the peculiar splendor with which
it shines among the stars.

a. **Name and Sign.**—This planet doubtless received its name on
account of its superior magnitude and splendor, Jupiter, or Jove, in the
ancient mythology, being the king of the gods. Its sign is supposed
to be an altered Z, the first letter of *Zeus*, the name of Jupiter among
the Greeks.

246. The aphelion distance of Jupiter is 498,500,000
miles; its perihelion distance 453,000,000 miles; hence its
mean distance is 475,750,000.

247. The ECCENTRICITY of its orbit is therefore 22,750,-

---

000 miles, or about .048 of its mean distance; being relatively nearly three times that of the earth.

248. The inclination of its orbit is very small, being only about 1° 19'.

249. Its SYNODIC PERIOD is found by observation to be about 399 days; hence its SIDEREAL PERIOD is about 4332 days, or 11ʸ 315ᵈ.

a. For 399ᵈ ÷ 365¼ᵈ = 1.0921 revolution performed by the earth during the synodic period; hence, Jupiter performs only .0921 of a revolution in 399ᵈ ; and 399ᵈ ÷ .0921 = 4332ᵈ +.

250. The APPARENT DIAMETER of Jupiter in opposition, when greatest, is about 50"; in conjunction, 31"; and at its mean distance is about 37". Its real equatorial diameter is about 87,500 miles.

a. For the least distance of Jupiter from the earth is 453,000,000 — 93,000,000 = 360,000,000 mile; and the maximum apparent semi-diameter is 25", the sine of which is .0001215, and 360,000,000 × .0001215 = 43,750 miles, which is the greatest, or equatorial, semi-diameter. Hence the diameter is 87,500 miles.

251. The OBLATENESS of Jupiter is very great, being about $\frac{1}{17}$ of its mean diameter, or about 5,000 miles.

a. Its polar diameter is therefore about 82,500 miles, and its mean diameter 85,000 miles. This remarkable degree of ellipticity in its figure is caused by the rapid rotation on its axis, which is performed in a little less than 10 hours (9ʰ 55½ᵐ); so that a point on the equator of this planet moves with a diurnal velocity of nearly 28,000 miles an hour, or 27 times as fast as at the earth.

b. It has already been explained (129 *a*) that the oblateness of a planet's figure results from the action of the centrifugal force; and it will be obvious that the degree of oblateness must depend on the *velocity of rotation* and the *density of the body*. It, in fact, varies directly as the square of the former, and inversely as the density, or the attraction of the body on its own matter. Thus, the velocity of Jupiter is about $2\frac{2}{5}$ times as great as the earth's [$24^h \div 10^h = 2\frac{2}{5}$], but its density is only $\frac{1}{4}$ as great, and $(2\frac{2}{5})^2 \times 4 = 23 +$. Therefore, if its density were as uniform as that of the earth, its oblateness would be $\frac{1}{299} \times 23 = \frac{1}{13}$ of its diameter. The internal parts of Jupiter must therefore be very much more dense than the external, the latter probably being considerably lighter than water.

Fig. 104.

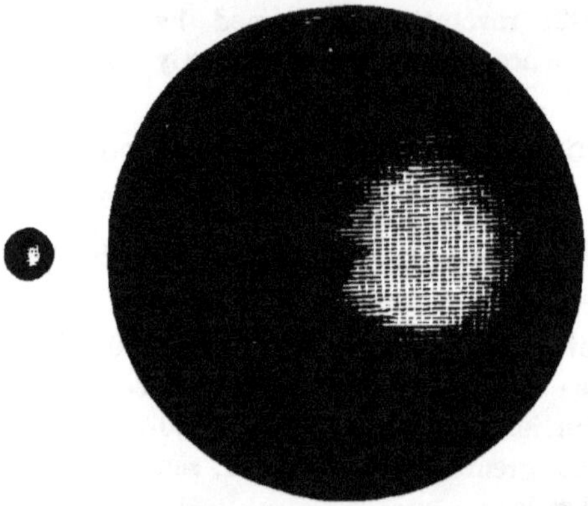

COMPARATIVE MAGNITUDES OF THE EARTH AND JUPITER.

c. **Volume, Mass, and Superficial Gravity.**—The volume of Jupiter is about 1244 times as great as the earth's; for $85{,}000 \div 7912 = 10.75$, and $(10.75)^3 = 1244$ (nearly). Its mass being only 301, its density is $301 \div 1244 = .242$, or nearly $\frac{1}{4}$. The force of gravity at the surface of the planet must therefore be $301 \div (10.75)^2 = 2.6 +$. So that a body weighing 1 lb at the earth's surface would weigh 2.6 lbs at Jupiter's; and since a body falls through 16 feet in the first second of time at the earth's surface, it would fall more than 41 feet at that of Jupiter.

d. **Orbital Velocity.**—This body, so inconceivably vast, is flying in its orbit with the velocity of 28,700 miles an hour, or nearly 500 miles a minute—a speed sixty times a great as that of a cannon ball. How tremendous is the exhibition of force here displayed!

252. The inclination of Jupiter's axis to that of its orbit is only 3° (3° 6′), much too small to cause any considerable change of seasons.

a. The phenomena of constant day and night take place, therefore, only within two small circles extending 3° from the poles. Exactly at the poles, the day and night are alternately about six years long. The cold at these parts of Jupiter must be intense beyond any that we can

QUESTIONS.—*c.* Volume, mass, and superficial gravity? *d.* Orbital velocity? 252. Inclination of its axis? *a.* What results from this?

conceive. The long absence of the sun, and its never rising more than 3° above the horizon, joined with the immense distance of the planet from that luminary, must all combine to intensify this rigor.

b. **Solar Light and Heat.**—The light and heat of the sun at Jupiter must be less than at the earth, in the inverse ratio of $(475,500,000)^2$ to $(90,000,000)^2$, or of 27 to 1. This feeble supply of light and heat may, however, be compensated by a greater density of the atmosphere, a higher calorific or luminous capacity of the soil, or a greater amount of internal heat than that possessed by the earth.

253. BELTS OF JUPITER.—When examined with a telescope the disc of Jupiter appears crossed by dusky streaks or belts, parallel to its equator, their general direction always remaining the same, although they constantly vary in number, breadth, and situation on the disc. Sometimes the disc is almost covered with them; while at others scarcely any are visible.

254. These dusky bands or belts are supposed to be the body of the planet seen between the clouds that constantly float in its atmosphere, and are thrown into zones or belts by the great velocity of its rotation. The cloudy zones are more luminous than the surface of the planet, on account of their more powerful reflection of the solar light.

a. The belts are not equally conspicuous, there being two generally which are more distinctly observable than others, and more permanent. These are situated, one on each side of the equator, and are separated by a clear space somewhat more luminous than the other parts of the disc. Toward the poles they are narrower and less dark; and they imperceptibly fade away a short distance from the eastern and western edges of the disc,—a phenomenon due, evidently, to the thickness of the atmosphere at those parts. Dark spots are also occasionally seen in connection with the belts.

In Fig. 105 are given two telescopic views of this planet; the first, from a drawing by Sir John Herschel, as it appeared September 23d, 1832; the

second, by Mädler, in 1834. The two dark spots shown in the latter were
employed to determine the time of the planet's rotation.

Fig. 105.

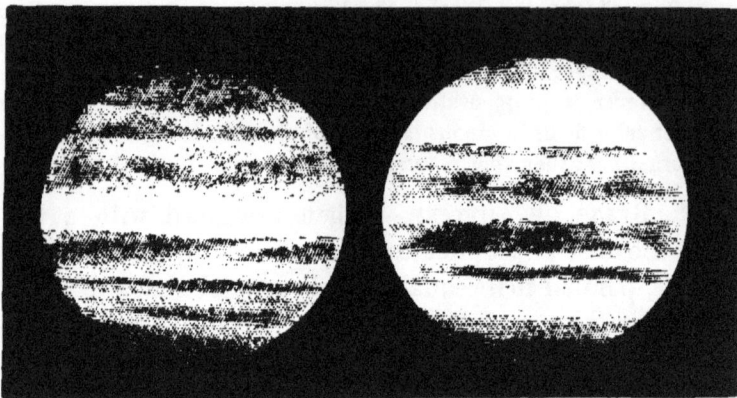

TELESCOPIC VIEWS OF JUPITER.

b. Notwithstanding the cloudy masses with which the atmosphere
appears to be charged, it is not thought that the latter has any very
great height above the planet's surface; for if such were the case the
edges of the disc, instead of being sharply defined as we see them,
would be nebulous and indistinct.

SATELLITES OF JUPITER.

255. The four satellites of Jupiter are among the most
interesting bodies of the solar system. They were first seen
by Galileo, in 1610, a short time after the invention of the
telescope, and were perceived to be satellites by their apparent movements with respect to the planet, alternately
approaching it, passing behind it, and receding from it;
sometimes also passing over its disc and casting their
shadows upon it.

a. These planets have been distinguished by particular names,

QUESTIONS.—*b.* Atmosphere? 255. Satellites—by whom discovered? Their apparent motions? *a.* How designated?

but are more generally designated by the numerals I., II., III., IV., according to their order from Jupiter

256. Their PERIODIC TIMES are, respectively, $1^d 18^h$; 3^d 13^h ; $7^d 4^h$; and $16^d 16^h$. The longest, it will bo seen, is but a little more than half that of the moon.

a. It will also be perceived that the second is *very nearly* twice the first ; and the third, twice the second.

257. Their diameters in approximate numbers, arc I., 2,800 miles ; II., 2,070 miles ; III., 3,400 miles ; IV., 2,900 miles ; all, excepting the second, being larger than the moon

a. These figures are based upon the measurements of their discs and a comparison of their apparent diameters with that of the planet as seen simultaneously. Thus, suppose the apparent diameter of Jupiter in opposition is found to be 45″, and the third satellite is measured at $1\frac{4}{5}″$; the diameter of the satellite must then be $\frac{1}{25}$ that of the primary planet, and $85,000 \times \frac{1}{25} = 3,400$.

b. As seen from Jupiter these bodies present quite large discs ; the apparent diameter of I. being 36′ ; of II., 19′ ; of III., 18′ ; and of IV., 9′. The first is therefore somewhat larger in appearance than that of the moon. The firmament of Jupiter must present a very beautiful diversity of phenomena. These various moons, all of which are occasionally above the horizon at one time, go through their phases within a few days; the first within 42 hours. To an inhabitant of the first satellite, the apparent diameter of Jupiter must be 19° ; that is, about 36 times as great as the moon ; while the amount of illuminated surface presented by it must be nearly 1300 times as great.

c. Although their volumes are quite large, their masses are very inconsiderable, owing to their very small densities, which are I., $\frac{1}{50}$; II., $\frac{1}{44}$; III., $\frac{1}{15}$; IV., $\frac{1}{18}$, the earth being 1. All are, thus, considerably lighter than water, and the first very much lighter than cork.

258. Their DISTANCES from Jupiter are, respectively, 264,000 miles, 423,000 miles, 678,000 miles, and 1,188,000 miles.

a. These are found by measuring their greatest elongations from the planet, and comparing these with its apparent diameter. Thus, the greatest elongations are respectively, 136″, 217″, 349″, and 611″; the apparent equatorial diameter of the planet being 45″. Dividing each elongation by 45″, we find the ratio to the planet's diameter of the distances of the satellites respectively. These are nearly I., 3; II., 4.8; III., 7.7; IV., 13.6. Fig. 106 shows the comparative extent of these elongations.

Fig. 106.

JUPITER AND ITS SATELLITES AT THEIR GREATEST ELONGATIONS.

b. The entire system of Jupiter is thus comprehended within a circular space of less than 2½ millions of miles in diameter, and subtends at its distance from the earth an angle less than 22′, or about ⅔ the apparent diameter of the moon. A telescope, the field of view of which would include one-half the area of the moon's disc, would exhibit Jupiter and all his satellites, as represented in Fig. 106.

c. A comparison of the periodic times and distances as above given will prove that they agree with Kepler's third law. Thus, taking I. and II. as an example, we find $(\frac{85}{42})^2 = 4.1$ (nearly), and $(\frac{423}{264})^3 = 4.1$ (nearly); hence, $(85)^2 : (42)^2 :: (423,000)^3 : (264,000)^3$.

259. The ORBITS of these bodies are almost circular, and very nearly in the plane of the planet's equator. They therefore make only a very small angle with the plane of its orbit (about 3°).

260. The ECLIPSES, OCCULTATIONS, and TRANSITS of the satellites present an endless series of interesting and useful phenomena; and the situation of their orbits causes them to occur with very great frequency.

a. I., II., and III. are eclipsed at every revolution: but so peculiarly related to each other are their motions that their simultaneous eclipse is impossible. Laplace demonstrated that the mean longitude of I., plus twice that of III., minus three times that of II., is always equal to

QUESTIONS.—*b.* Angular space covered by the system? *c.* Kepler's law—how applied? 259. Figure and position of the orbits? 260. Eclipses? Why frequent? *a.* How many eclipses may occur at Jupiter during a Jovian year?

180°. Hence, when two are eclipsed, the other must be on the opposite side of the planet. This is called the *libration of the satellites.* All four are, however, occasionally invisible, being concealed either behind or in front of the planet. This occurred last in August, 1867. It has been computed that, during a year of Jupiter, an inhabitant of the planet might behold 4,500 solar and lunar eclipses.

b. During the transits the satellites appear like *bright* spots passing from east to west across the disc, preceded or followed by their shadows, which seem like small round dots as black as ink.

Fig. 107.

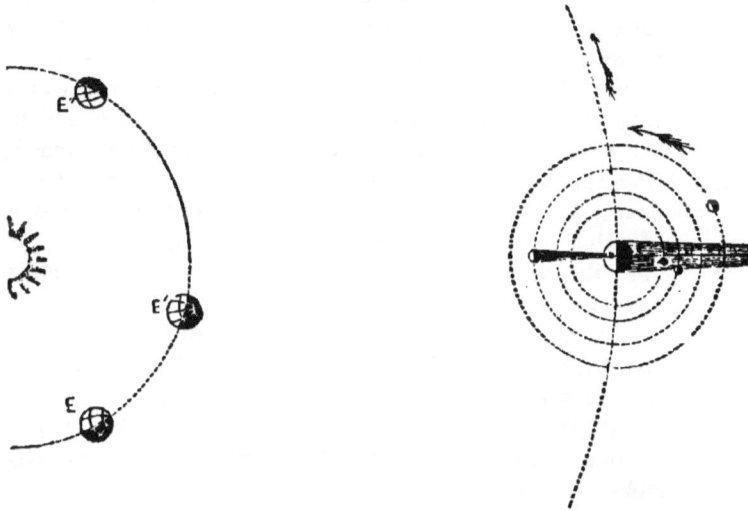

ECLIPSES, OCCULTATIONS, AND TRANSITS OF JUPITER'S SATELLITES.

In Fig. 107, to an observer at E, I. is represented as eclipsed; II., as just passing into the shadow of the planet; III., just before a transit, the shadow *preceding;* and IV., at the point of occultation. At E', I. has just passed behind the disc; II. is in occultation; III., a transit, both shadow and satellite being on the disc, the shadow preceding; IV., just emerging from behind the planet; at E'', I. and II. are behind the disc, III. is in transit, but the shadow follows the satellite; IV., just after an eclipse.

261. Since the occurrence of these eclipses can be exactly predicted, they serve to mark points of absolute time; so that if the precise moment at which they will occur at any

particular place has been computed, and the actual time of
their occurrence at any other place is noted, a comparison
of the two will give the difference of time, and, of course,
the difference of longitude, between the two places.

a. Thus, if a mariner perceives, by the nautical almanac, that the
eclipse of a satellite will occur at 9 o'clock P.M., Washington time,
and he notices that the eclipse does not take place till 11 o'clock P.M.,
he can infer that his position is 2 hours, or 30°, east of Washington.

b. **Velocity of Light found by the Eclipses of Jupiter's Satellites.**—In the prediction of these eclipses, a constant variation was for
several years found to exist between the calculated and observed time
of the occurrence, with this additional fact, that the eclipse was later
as Jupiter receded from the earth and earlier as it approached the
earth ; being about 16ᵐ 35½ˢ earlier in opposition than in conjunction.
These observations were made by Olaus Roemer, a Danish astronomer;
and in 1675 he promulgated the theory, to account for the phenomena,
that the passage of light from a luminous body is not instantaneous,
but moves with a certain definite but immense velocity, requiring 16ᵐ
35½ˢ to cross the earth's orbit. This theory has been universally
accepted, and certain experiments recently made in France, by M.
Fizeau and others, have confirmed it. The velocity of light must
therefore be 184,000 miles a second. For 183,000,000 miles (distance
across the earth's orbit) divided by 995½ (number of seconds in 16ᵐ
35½ˢ), gives 184,000 (nearly). Light must therefore require 8¼ minutes to pass from the sun to the earth. So great a velocity is entirely
inconceivable.

III. SATURN ♄

262. SATURN, the second of the major planets, is the
centre of a very large and peculiar system, being attended
by eight satellites and encompassed by several rings. It
shines with a dull yellowish light.

a. **Name and Sign.**—Saturn, in the ancient mythology, was one of
the older deities, and presided over time, the seasons, etc. He was
represented as a very old man carrying a scythe in one hand. The
sign of the planet is a rude representation of a scythe.

QUESTIONS.—*a.* Illustration? *b.* What important discovery made by Roemer? In
what way? What is the velocity of light? 262. General description of Saturn? *a.*
Name and sign?

263. The aphelion distance of Saturn is about 921 millions of miles; the perihelion distance, 823 millions; the mean distance being therefore 872 millions.

a. This is nearly twice the distance of Jupiter, between which and Saturn there is a vast space of nearly 400 millions of miles, in linear breadth, through which there rolls no planetary body. Light requires about 1¼ʰ to pass from the sun to Saturn.

264. The ECCENTRICITY of Saturn's orbit is nearly 50 millions of miles, or about .056 of its mean distance, being but little greater, relatively, than that of Jupiter.

265. The INCLINATION OF ITS ORBIT to the plane of the ecliptic is about $2\frac{1}{2}°$ (2° 29' 36'').

266. Its SYNODIC PERIOD is 378 days (378.07ᵈ), and its SIDEREAL PERIOD, 10,759 days or about $29\frac{1}{2}$ years.

a. For 378.07ᵈ ÷ 365.25ᵈ = 1.03514; and 378.07ᵈ + .03514 = 10759ᵈ (nearly) The year of Saturn contains, therefore, about 25,000 of its own days.

267. The greatest apparent diameter of Saturn is 21''; its least, $14\frac{1}{2}$''. Its real equatorial diameter is about 74,000 miles.

a. For the least distance from the earth is 823 millions of miles — 93 millions = 730 millions : and the sine of $10\frac{1}{2}$'' is about .00005068, which being multiplied by 730,000,000 will give 37,000 (nearly)—the semi-diameter.

268. The OBLATENESS of Saturn is greater than that of any other planet, being a little more than $\frac{1}{10}$ of its equatorial diameter, or 7,800 miles.

a. Hence its polar diameter is only 66,200 miles : its mean diameter being 70,100 miles.

269. The AXIAL ROTATION is performed in about $10\frac{1}{2}$ hours (10ʰ 29ᵐ 17ˢ).

a. This was the determination reached by Sir William Herschel by means of observations made on the belts which, like those of Jupiter, cross the planet's disc. Subsequent observations have indicated but little variation from it.

b. The equatorial velocity of Saturn is, therefore, more than 22,000 miles an hour; and as its density is very small, being only $\frac{1}{8}$ of the earth's, its oblateness should be, according to the law stated in Art. 251, *b,* $\left(\dfrac{24}{10\frac{1}{2}}\right)^2 \times \frac{2\frac{1}{4}} = 40\frac{1}{2}$; that is, $40\frac{1}{2}$ times as great as the earth's. But $\frac{1}{299} \times 40\frac{1}{2} = \frac{81}{299} = .1355$ (nearly), or about $\frac{1}{8}$. So that its observed oblateness is much less than it ought to be in accordance with this law. The measurement of Saturn's apparent diameter is, however, so difficult, in consequence of the rings, that there may be considerable error in the statement of its oblateness given above.

c. Volume, Mass, and Density.—The volume of Saturn as compared with that of the earth, is $(\frac{71100}{7977})^3 = 695\frac{1}{2}$, and its mass has been found to be 90; hence its density is (as above stated), $90 \div 695\frac{1}{2} = \frac{1}{8}$ (nearly), the earth's being 1; or $5\frac{1}{2} \times \frac{1}{8} = .726$ as compared with water; that is, somewhat lighter than oak wood.

d. Superficial Gravity.—This must be, according to the figures above given, $(\frac{7977}{71100})^2 \times 90 = 1.15$ (nearly). Hence, a body at the surface of Saturn weighs only about $\frac{1}{7}$ more than at the surface of the earth, notwithstanding the immense size of that planet.

270. The INCLINATION OF ITS AXIS toward the plane of its orbit is about 27° (26° 48′ 40″), or a little greater than that of the earth.

a. That is, its axis makes an angle of 27° with a *perpendicular* to its orbit. The angle which it makes with the plane of its orbit is 90° — 27° = 63° The position of the axis is such that its inclination toward the plane of the ecliptic is about 28° 10′; and like that of the earth and those of the other planets, as far as it has been ascertained, the axis remains parallel to itself during the orbital motion.

b. The Seasons of Saturn must therefore be similar to those of the earth, but like the year, $29\frac{1}{2}$ times as long.

c. Solar Light and Heat.—The distance of Saturn from the sun being more than $9\frac{1}{2}$ times as great as the earth's, the apparent diameter of the sun must be less in the same proportion, or $32' \div 9\frac{1}{2} = 3'\ 22''$. Hence the light and heat, considered with reference to distance, must be, compared with the earth's, as $(202'')^2$ to $(1920'')^2$, or as 1 to $(9.53)^2$; that is, only $\frac{1}{91}$ of the earth's.

271. SATURN'S BELTS.—This planet, when viewed with a good telescope, appears to be encompassed with dusky belts; but they are far more indistinct than those of Jupiter; and instead of crossing the disc in straight lines like those of that body, they generally present a curved appearance,—an indication of the axial inclination.

a. Sir William Herschel inferred the existence of a dense atmosphere surrounding Saturn, both from the changes constantly occurring in the number and appearance of the belts, and the appearance of the satellites at the occurrence of occultations. The nearest was observed to cling to the edge of the disc about twenty minutes longer than would have been possible had there been no atmosphere to refract the light. Indications of accumulations of ice and snow at the poles have also been detected, similar to those of Mars.

272. RINGS.—Saturn is encompassed by three or more thin, flat rings, all situated exactly or very nearly in the plane of its equator.

a. **History of their Discovery.**—In his first telescopic examination of this planet, Galileo noticed something peculiar in its form. As seen through his imperfect instrument, it appeared to him to have a small planet on each side; and hence, he announced to Kepler the curious discovery that " Saturn was threefold;" but continuing his observations, he saw, to his great astonishment, these companion bodies (as he thought) grow less and less, and finally disappear. For fifty years afterward the true cause of the appearance remained un- known, the distortion of the planet's form being supposed to arise from two *handles* attached to it. Hence they were called *ansæ,* the Latin word for handles. Huyghens, in 1659, discovered the real cause of the phenomenon, and announced it in these words : " The planet is sur- rounded by a slender, flat ring, everywhere distinct from its surface, and inclined to the ecliptic." The division of the ring into two was discovered by an English astronomer in 1665. We have now certain knowledge of the existence of three rings, and some indications of several more.

273. Two of these rings are very distinctly observed, and are designated the *interior* and the *exterior ring*. The former is about 16,500 miles wide; the latter, 10,000 miles. The distance of the interior ring from the planet is about 18,350 miles; and the interval between these rings, about 1,700 miles. The thickness of the rings does not exceed 250 miles, and may be much less.

a. The diameter of the exterior ring, at the mean distance of the planet, subtends an angle of 40″, and is, consequently, nearly 170,000 miles. It will thus be evident that 1″ of angular space at the distance of Saturn is equal to nearly 4,250 miles; so that if the ring were 250 miles in thickness, it would subtend only about $\frac{1}{17}$″. The difficulty of determining this precisely will at once be obvious. The mass of the ring has been computed to be equal to the 118th part of the planet's mass, from its effect in disturbing some of the satellites; and this would prove, if its density is equal to that of the planet, that its thickness is about 140 miles.

274. Within the interior ring there is a dusky or semi-transparent ring, having a crape-like appearance as it stretches across the bright disc of the planet. (See Fig. 108.)

a. None of the early observers noticed this.. In 1838, the Prussian astronomer, Galle, perceived a gradual shading off of the interior ring toward the planet. His announcement of the fact elicited no attention until, in 1850, the distinguished astronomer of our own country, G. P. Bond, plainly discovered and announced (Nov. 11) the existence of this dusky ring: before, however, the intelligence had reached England, the discovery had been made (Nov. 18) there also, by the celebrated observer, Dawes. The transparency of this ring was fully established in 1852 by Dawes and Lassell. There are also very decided indications that this dark ring is also double.

Fig. 108 shows the planet as seen by Dawes in 1852. The form and partial transparency of the dark ring are clearly represented; the interval

Fig 108.

TELESCOPIC VIEW OF SATURN.—*Dawes.*

between the interior and exterior rings is also visible, as well as the line in the latter, supposed to indicate another division of the rings.

b. **Rotation of the Rings.**—It was discovered by Sir William Herschel, by observing certain bright spots seen on the surface of the rings, that they rotated on an axis perpendicular to their plane, and very nearly coincident with that of the planet. The time of rotation is about 10^h 32^m, which is the time required by a satellite situated at a distance from the planet equal to the centre of the rings, to perform its revolution, according to Kepler's third law. As the planet revolves around the sun, the rings constantly remain parallel to themselves.

c. **Stability of the Rings.**—Observations of great delicacy have shown that the rings are not exactly concentric with the planet, the centre of gravity of the former revolving in a small orbit round that of the latter; and Laplace showed that this is an essential condition of their stability; since, if they were precisely concentric, a very slight disturbance, such as the attraction of a satellite, would be sufficient to destroy their equilibrium and finally precipitate them upon the planet.

d. **Physical Constitution of the Rings.**—The rings are evidently

QUESTIONS.—*b.* Rotation of the rings? *c.* Are the rings concentric with the planet? *d.* Why have the rings been supposed to be fluid? What other hypothesis.

opaque, since they cast a shadow upon the planet and are in turn obscured by that of the planet itself; and it was, until quite recently, thought that they consisted of solid matter. This, however, is now generally considered to be at variance both with theory and observation; it being shown that the equilibrium of *solid* rings could not long be preserved, except by an arrangement which certainly does not exist. Moreover, several subordinate divisions have been remarked in the bright rings, portions of which are of a different shade, presenting the "appearance of four or five concentric and deepening bands," compared oy Lassell to the "steps of an ampitheatre." These shaded bands and the lines which separate them do not always present exactly the same appearance. The idea has therefore been entertained that the rings might be *fluid*, not only from the circumstances above enumeraied, but because minute observations disclose the fact that the rings have become broader and thinner than they were when first discovered. A more generally received hypothesis is, that the rings consist of vast numbers of satellites revolving around the planet; and that being more sparsely scattered in the dark ring, they reflect the light imperfectly and disclose the bright disc of the planet between them This hypothesis not only explains all the phenomena of the rings, but is consistent with other phenomena presented by the solar system, to be referred to hereafter.

275. APPEARANCE OF THE RINGS.—The rings, although circular, appear like ellipses because, being inclined to the plane of the ecliptic, they are always viewed obliquely. They become invisible when the dark side is turned toward the earth; and, when its edge only is presented, are seen, in very powerful telescopes, as a mere thread of light, cutting the disc of the planet. Sometimes the satellites appear along this thread like a series of brilliant beads on a string.

a. The edge only of the rings is seen when their plane if prolonge would pass through the earth; also when the same plane pass through the sun, so that the edge only is illuminated. In each these cases, the rings must disappear or present only a thread light. They disappear also when their unillumined side is turne toward the earth. which must, of course, occur when their plane passes between the earth and sun, so that the rays of the latter fall only on that side of the rings which is turned away from the earth.

Fig. 109.

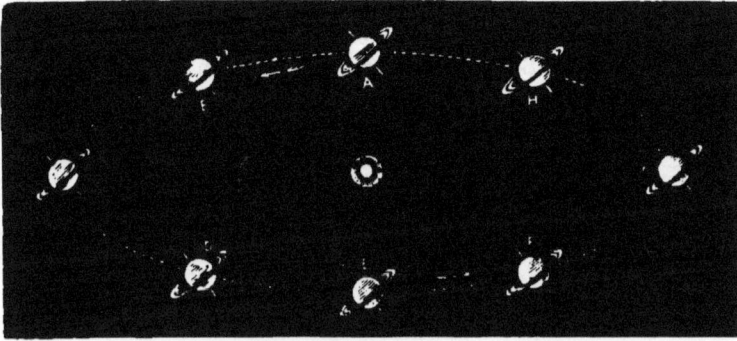

SATURN IN DIFFERENT PARTS OF ITS ORBIT.

Fig. 109 represents Saturn in different parts of its orbit, the direction of the axis and the position of the plane of the rings constantly remaining the same. At A or E, the time of the planet's equinox, the plane of the rings passes through the sun, so that its edge only is illuminated, wherever the earth may be situated, which is to be conceived as revolving in a small orbit within that of Saturn. At the solstice C, the southern side of the rings is exhibited; and at G, the northern side. At the intermediate points the rings are viewed obliquely. It will be obvious that, owing to the comparatively small size of the earth's orbit, the plane of the rings can pass through the earth, or between it and the sun, only a short time before or after the equinox, and, as this must occur at each equinox, that the disappearance of the rings from this cause must occur twice during each sidereal revolution of the planet, or at intervals of 14¾ years. The last disappearance took place in 1862, the next will occur in 1877. At the present time (1867), the northern surface of the rings is visible.

276. SATELLITES.—Saturn is attended by eight satellites, seven of which revolve very nearly in the plane of its equator, the orbit of the eighth, or most distant satellite, making with that plane an angle of 12¼°.

a. Names.—The names of these satellites in the order of their distances from Saturn, beginning with the nearest, are the following:—

1. *Mimas*, 2. *Enceladus*, 3. *Tethys*, 4. *Dio'ne*, 5. *Rhea*, 6. *Titan*, 7. *Hy perion*, and 8. *Jap'etus*.

b. **History of their Discovery.**—Titan was discovered by Huy. ghens, in 1665; Tethys, Dione, Rhea, and Japetus, by Cassini, within about twenty years afterward; Mimas and Enceladus, by Sir William Herschel, in 1787 and 1789; and Hyperion was discovered by Lassell, at Liverpool, and by Bond, at Cambridge, Mass., on the *same evening,* September 19, 1848.

c. The following are their periods and distances from the primary:

	PERIODS.	DISTANCES.		PERIODS.	DISTANCES.
1. MIMAS	22½ ʰ	121,000	5. RHEA	4 12½ ʰ	843,000
2. ENCELADUS	1ᵈ 9ʰ	155,000	6. TITAN	15 23 ʰ	796,000
8. TETHYS	1ᵈ 21ʰ	191,000	7. HYPERION	21ᵈ 7ʰ	1,006,000
4. DIONE	2 18 ʰ	246,000	8. JAPETUS	79ᵈ 8ʰ	2,513,000

277. The largest of the satellites is Titan, its diameter being 3,300 miles, which is larger than that of Mercury. The sizes of the others are very much less.

a. That of Japetus is 1,800 miles; Rhea, 1,200; Mimas; 1,000 Tethys and Dione, 500; Enceladus and Hyperion, unknown.

b. The orbit of Japetus subtends an angle of only 21⅓′; so that this magnificent system of Saturn with his rings and eight satellites, at its immense distance from the earth, is contained within a space in the heavens less than one-half the disc of the moon.

c. In 1862, while the ring was invisible, the rare phenomenon occurred of a transit of Titan across the disc of the primary. Tʰ shadow was observed by Dawes and others. The same phenome‑ was observed by Sir William Herschel in 1789.

d. The variations in the light of Titan indicated to Sir V Herschel an axial rotation of the satellite, which, like that of satellites whose periods have been discovered, is performed in time as the revolution around the primary.

278. The CELESTIAL PHENOMENA at Saturn mᵣ

a scene of extreme beauty and grandeur. The starry vault, besides being diversified by so many satellites, presenting every variety of phase, must be spanned, in certain parts of the planet, and during different portions of its long year, by broad, luminous arches, extending to different elevations, according to the place of the observer, and receiving upon their central parts the shadow of the planet.

IV. URANUS. ♅

279. URANUS was discovered in 1781 by Sir William Herschel. It shines with a pale and faint light, and to the unassisted eye is scarcely distinguishable from the smallest of the visible stars.

a. **History of its Discovery.**—This planet had been observed by several astronomers previous to its discovery by Herschel, but had been mapped as a star at least twenty times between 1690 and 1771, its planetary character not having been discerned ; and even Herschel, on noticing that its appearance was different from that of a star, was not aware that he had discovered a new planet, but supposed it to be a comet, and so announced it to the world, April 19th, 1781. It was, however, in a few months, evident that the body was moving in an orbit much too circular for a comet ; but its planetary character, sug-
ﬞﬞﬞﬞsted first by Lexell, in June, 1781, was not fully established until
ₙₙₙ, when Laplace partly calculated the elements of its orbit. This,
ﬞﬞer, does not detract from the merit of Herschel, in making this
ﬞﬞﬞy ; for, the attention of astronomers having been called to this
ﬞﬞ one of a peculiar character, and not *sidereal*, it was a simple
ﬞﬞﬞetermine whether it was a planet or a comet. The merit of
ﬞﬞﬞry consisted in that delicacy of observation, that skill in the
ﬞﬞuments, and, more than all, that unfailing perseverance
ﬞﬞﬞterized Herschel, and made him the great astronomer of

ﬞnd **Sign.**—Herschel proposed to call the new planet

"Georgium Sidus," *George's Star*, in compliment to his friend and patron, King George III. This name not being accepted by foreign astronomers, Lalande proposed to name it "Herschel," after its great discoverer; and by this designation it was, for some time, quite generally known. The scientific world has now definitely settled upon the name, suggested by Bodé, of *Uranus*, which, in the Grecian mythology, was the name of the oldest of the deities, the father of Saturn, as Saturn was the father of Jupiter. The name of the discoverer is, however, partly connected with the planet by the sign, which is the letter H with a suspended orb.

280. The aphelion distance of Uranus is about 1,836 millions of miles; its perihelion distance, 1,672 millions; the mean distance being 1,754 millions, which is more than 19 times (19.183) that of the earth.

a. Light requires 2 hours $33\frac{1}{2}$ minutes to pass from the sun to Uranus; for $8^m \times 19.183 = 153.464^m = 2^h 33\frac{1}{2}^m$ (nearly). Sunrise and sunset are therefore not perceived by the inhabitants of Uranus for two hours and a half after they really occur, for the light which proceeds from the sun when it touches the plane of the horizon does not reach the eye until $2\frac{1}{2}$ hours afterward.

b. The distance of this planet from the sun is so vast that the greatest elongation of the earth as seen from it is only about 2°. That of Jupiter is only $16\frac{1}{2}$°, while its apparent diameter is but little greater than that of Mercury as seen from the earth. Even Saturn departs only about 29' from the sun, its apparent diameter being less than 20". The inhabitants of the planet, if any there be, must therefore possess much less opportunity than ourselves to become acquainted with the constituent members of the great system to which they belong.

281. The ECCENTRICITY of the orbit of Uranus is about 82 millions of miles, or about .047 of its mean distance.

282. The INCLINATION OF ITS ORBIT is less than that of any other planet, being only $46\frac{1}{4}'$.

QUESTIONS.—280. What is its distance from the sun? *n.* What time does light require to pass from the sun to Uranus? Effect on apparent sunrise and sunset? *b.* Elongations and apparent diameters of the planets as seen from Uranus? 281. Eccentricity of its orbit? 282. Inclination of its orbit?

a. Nevertheless, so vast is its distance that, at its greatest latitude, it may depart from the plane of the ecliptic more than 24 millions of miles.

283. Its SYNODIC PERIOD is 369.65 days; and therefore its sidereal period is 30,687 days, or about 84 years.

a. The computation may be made as in the case of the other planets: 369.65 ÷ 365.25 = 1.012046 +; that is, Uranus performs about .012046 of a sidereal revolution during the synodic period. Hence, 369.65ᵈ ÷ .012046 = 30,687ᵈ (nearly) is the sidereal period.

b. In the case of a very distant planet, the sidereal period may be readily found by observing the daily arc of movement of the planet when in quadrature: for, at that time, the line joining the earth and planet is a tangent to the earth's orbit (see Fig. 28), so that, for a short time, the earth moves either toward or from the planet, and does not affect the apparent motion of the latter, while its distance is so great that its geocentric increase in longitude is almost equal to its heliocentric. Now, the apparent daily increase of the longitude of Uranus in quadrature is 42.23″, and 360° ÷ 42.23″ = 30,689, which gives a near approximation to the true sidereal period.

284. The greatest apparent diameter of Uranus is about 4″; and as 1″, at the least distance of this planet, subtends 8,350 miles, the real diameter must be 33,400 miles. (By more exact calculations, it is found to be 33,247 miles.)

a. The *oblateness* has not positively been ascertained. Mädler estimates it to be as much as ¹⁄₁₀. The *volume* of Uranus is about 72½ times that of the earth; but its *mass* is only 13 times; hence its *density* is less than ⅓ that of the earth, or about equal to that of water.

285. As the disc of Uranus presents neither belts nor spots, the period of its rotation and its axial inclination still remain unknown. It is thought, from the positions of the orbits of the satellites, that the inclination of its axis is

made with the best instruments to detect the others. In 1847 two others, situated within the orbit of the nearest discovered by Herschel, were detected, one by Lassell and the other by O. Struve.

b. The following are the names of these satellites, with their periods and distances :

	PERIODS.	DISTANCES.		PERIODS.	DISTANCES.
1. ARIEL	2d 12¾h	123,000	3. TITANIA	8d 17h	281,000
2. UMBRIEL	4d 3¼h	171,000	4. OBERON	13d 17h	376,000

c. Their orbits are inclined to the plane of that of the primary at an angle of 79° ; but, as their motion is retrograde, it seems probable that the poles have been reversed in position, the south pole being north of the ecliptic, and *vice versa.* The inclination is properly, therefore, 101°.

V. NEPTUNE ♆

287. NEPTUNE is the most distant planet known to belong to the solar system. It was first observed in 1846 by Dr. Galle at Berlin ; but its existence had been predicted, and its position in the heavens very nearly ascertained by the calculations of M. Leverrier, in France, and Mr. Adams, in England ; these calculations being based upon certain observed irregularities in the motion of Uranus.

QUESTIONS.—286. How many satellites attend Uranus? Direction of their orbital motion? *a.* History of their discovery? *b.* Names, periods, and distances? *c.* Position of their orbits and poles ? 287. By whom, and how was Neptune discovered?

a. **History of its Discovery.**—The discovery of this planet was one of the proudest achievements of mathematical science in its application to astronomy, and afforded a more striking proof of the truth of the great law of universal gravitation than had previously been ascertained. After the discovery of Uranus, in 1781, it was ascertained that the planet had several times been observed by astronomers, and its place recorded as a star. These positions of the planet could not, however, be reconciled with those recorded after its actual discovery; and observation soon showed that its motion was at certain points increased, and at others diminished, by some force acting beyond it and in the plane of its orbit. These facts suggested the existence of another planet, revolving in an orbit exterior to that of Uranus, and, according to Bodé's law, extending nearly twice as far from the sun. Adams and Leverrier almost simultaneously undertook to find, by mathematical analysis, where this planet must be in order to produce these perturbations. The former reached the solution of this wonderful problem first, and, in October, 1845, after three years of toil, communicated to Mr. Airy, Astronomer Royal, the result, pointing out the position of the planet and the elements of its orbit. The search for the planet was not, however, commenced until Leverrier published the result of his labors, which was found to agree so closely with that attained by Adams, that astronomers both in France and England prepared to construct maps of the part of the heavens indicated, in order to detect the planet.

In this they were anticipated by the Berlin observer, who, being informed by Leverrier of the result of his computations, and having by a fortunate coincidence just received a newly prepared star-map of the 21st hour of right ascension (the part of the heavens designated by Leverrier), immediately compared it with the stars, and found one of them missing. The observations of the following evening, by detecting a retrograde motion of this star, established its true character. It was the planet sought for, and, wonderful to relate, was found only 52' from the place assigned by Leverrier. He had also stated its apparent diameter at 3.3''; it was found by actual measurement to be 3''. Adams's determination of the place of the supposed planet differed from the true place by about 2°.

b. **Name and Sign.**—This planet, according to the system of mythological designations, was, after considerable discussion, called

Neptune. The sign is the head of a trident—the peculiar symbol of this deity.

288. The APHELION DISTANCE of Neptune is 2,770 millions of miles; its perihelion distance, 2,722 millions; its mean distance being 2,746 millions.

a. This is about 30 times the distance of the earth; but according to Bodé's law, it should have been 38.8 times; so that this remarkable relation of the planets, failing in this instance, ceases to be a *law*, and becomes, apparently, only a curious *coincidence.*

b. So immense is the distance of Neptune that only Saturn and Uranus can be seen from it. If there are astronomers, however, on the planet, they must have much better opportunities than ourselves for becoming acquainted with the distances of the stars; since, at opposite periods of their long year, they are situated at positions in space about 5.500 millions of miles apart.

c. Since the distance of Neptune from the sun is 30 times that of the earth, light requires $8^m \times 30 = 4^h$, to reach that planet.

289. The ECCENTRICITY of the orbit of Neptune is about 24 millions of miles, which is only .0087 of its mean distance; so that it is, relatively, but little more than one-half that of the earth's orbit.

290. The INCLINATION OF ITS ORBIT to the plane of the ecliptic is very small, being only $1\frac{3}{4}°$ ($1° 47'$).

a. The sine of $1° 47'$ is .031; hence Neptune, when at its mean distance from the sun, and at the point of greatest latitude north or south of the ecliptic, must be more than 85 millions of miles from the plane of that circle; for, $2.746,000,000 \times .031 = 85,126,000.$

291. Its SYNODIC PERIOD is about $367\frac{1}{2}$ days (367.48234); hence its SIDEREAL PERIOD is 60,127 days, or about $164\frac{1}{4}$ years.

a. It is more difficult to calculate the sidereal periods of these

QUESTIONS.—288. Aphelion, perihelion, and mean distances? *a.* Does it agree with Bodé's law? *b.* Which planets can be seen at Neptune? *c.* How long does light require to pass from the sun to Neptune? 289. Eccentricity of its orbit? 290. Inclination? *a.* How far may it depart from the plane of the ecliptic? How is this calculated? 291. Synodic period? Sidereal period? *a.* How calculated?

remote planets; since the synodic period is so nearly equal to the sidereal period of the earth, that the fraction of a revolution performed during the latter is very small. In the case of Neptune it is a little over .0061118; that is, $367.48234^d + 365.25^d = 1.0061118 +$; and $367.48234^d + .0061118 = 60,127$ days (nearly).

292. The APPARENT DIAMETER of Neptune when greatest is 2.9''; hence its real diameter must be nearly 37,000 miles.

a. For the least distance of Neptune from the earth is 2,722 millions — 93 millions = 2,629 millions; now the sine of 2.9'' is .000014; and 2,629 millions multiplied by this small fraction will give 36,806 miles.

b. **Volume, Mass, and Density.**—The volume of Neptune, if calculated by the method previously explained, will be found to be very nearly 99 times as great as that of the earth, and consequently is only about $\frac{1}{14}$ as large as Jupiter. Its *mass* is nearly 17 times (16.76) as great as the earth's [Prof, Pierce]; consequently its density must be about $\frac{1}{6}$ that of the earth, or somewhat more than $\frac{2}{10}$ as heavy as water.

c. **Solar Light and Heat.**—The apparent diameter of the sun as seen at Neptune must be a little more than 1'; for, $32' \div 30.037$ (ratio of of Neptune's distance to the earth's)=64''(nearly). Hence, the sun at this planet looks but little larger than Venus; but its light is vastly more brilliant. For, since the intensity of light varies inversely as the square of the distance, and $(30.037)^2 = 902$ (nearly), the light at Neptune must be $\frac{1}{902}$ of that at the earth, and hence is nearly equal to that of 670 full moons (157, *b*). This is probably as great as that which would be produced by 20,000 stars shining at once in the firmament, each equal to Venus when its splendor is greatest.

293. A SATELLITE of this planet was discovered by Lassell in October, 1846, and was afterward observed by several other astronomers.

a. From observations made about the same time the existence of another satellite was suspected, as well as a ring analogous to that of Saturn; but the most diligent and careful scrutiny with very powerful telescopes has failed to detect any indications of the truth of these conjectures.

b. **Distance of the Satellite.**—The observations made by eminent astronomers (principally those of M. Struve, Mr. Lassell, and Mr. B)nd) have shown that the greatest elongation of the satellite from its primary is 18″, the apparent diameter of the latter being at the same time 2.8″. Hence its distance must be 18″ ÷ 2.8″ = 6⅔ diameters, or 12⅔ radii, of the planet: and 18,500 × 12⅔ = 238,000 miles, or about the same as the moon's distance from the earth.

c. **Inclination, Period, and Rotation.**—The orbit of this satellite is nearly circular, and is inclined to the orbit of Neptune in an angle of 29°. Its motion, like that of the satellites of Uranus, is retrograde, or from east to west Its sidereal period, as determined by Lassell at Malta, in 1352, is 5^d 21^h. Periodical changes in its brightness were observed by Lassell, which indicated that this satellite, like others in the system, rotates on its axis in the same time that it revolves around its primary.

d. **Are there Planets beyond Neptune?**—This is a question which we are at present entirely unable to answer. Future generations may, with greater resources of science and mechanical skill, disclose new marvels in our system, and detect other bodies obedient to the dominion of its great central sun. The nearest of the stars is known to be nearly 7,000 times as far from Neptune as that body is from the sun ; and it is by no means improbable, therefore, that so vast a space should contain planetary bodies reached by the solar attraction, but very far beyond the sphere of any other central luminary. It will require, however, far greater means than we possess to bring this to a practical determination.

QUESTIONS.—*b.* What is the distance of this satellite from the primary ? How cal culated ? *c.* Its inclination of orbit ? Orbital revolution—period and direction ? Axial rotation ? *d.* is Neptune the remotest planet ?

CHAPTER XIV.

294. The MINOR PLANETS are a large number of small bodies revolving around the sun between the orbits of Mars and Jupiter. The number discovered up to the present time (1869) is 106.

a. **Discovery of Ceres and Pallas.**—The existence of so large an interval between Mars and Jupiter, compared with the relative distances of the other planets, for a long time engaged the attention and incited the researches of astronomers. Kepler conjectured that a planet existed in this part of the system, too small to be detected; and this opinion received considerable support from the publication of Bodé's law in 1772. When Uranus was discovered, in 1781, and its distance was found to conform to this law, the German astronomers became so confident of the truth of this bold conjecture of Kepler, that, in 1800, they formed, under the leadership of Baron de Zach, an association of 24 observers to divide the zodiac into sections and make a thorough search for the supposed planet. This systematic exploration had, however, been scarcely commenced, when, in 1801, Piazzi, an Italian astronomer, while engaged in constructing a catalogue of stars, detected a new planet. It was called by him *Ceres*. In the next year, while looking for the new planet, Olbers discovered another, which he called *Pallas*.

b. **Discovery of Juno and Vesta—Theory of Olbers.**—The extre.ne minuteness of the new planets, and the near approach of their orbits at the nodes, led Olbers to suppose that they might be the fragments of a much larger planet once revolving in this part of the system, and shattered by some extraordinary convulsion. Believing

---

QUESTIONS.—294. What are the minor planets? Their number? *a*. How and by whom were Ceres and Pallas discovered? *b*. Juno and Vesta? Theory of Olbers?

that other fragments existed, and that they must pass near the nodes of those already found, he resolved to search carefully in the direction of those points; but while he was thus engaged, Harding, of the observatory of Lilienthal, discovered, in 1804, very near one of those points, a third planet, which he called *Juno*. Olbers, still further stimulated by this event to continue the investigation which he had commenced, was at length, in 1807, rewarded by discovering a fourth planet, *Vesta*, near the opposite node. From this date until 1845, no additional discovery was made. These small planets were called *Asteroids* by Herschel, from their resemblance, in appearance, to stars.

c. **Discovery of the other Minor Planets.**—In 1845, M. Hencke, an amateur astronomer of Driessen, after a series of observations continued for fifteen years with the use of the Berlin star-maps, discovered *Astræa*, the fifth of this singular zone of telescopic planets. The others have been discovered in the following order: In 1847, *Hebe, Iris, Flora;* 1848, *Metis;* 1849, *Hygeia;* 1850, *Parthen'ope, Victoria,* and *Egeria;* 1851, *Ire'ne* and *Eunomia;* 1852, *Psyche, Thetis, Melpom'enĕ, Fortu'nă, Massilia, Lutetia, Calli'opĕ,* and *Thalī'a;* 1853, *Thĕmis, Phoce'a Proser'pina,* and *Euter'pĕ;* 1854, *Bello'na, Amphitrī'tĕ, Urania, Euphrōs'ynĕ, Pomo'na,* and *Polyhym'nia;* 1855, *Circĕ, Leuco'thea. Atalan'ta,* and *Fi'dĕs;* 1856, *Le'da, Lætita, Harmonia, Daph'nĕ,* and *Isis;* 1857, *Ariad'nĕ, Nȳ'sa, Euge'nia, Hestia, Mel'etĕ, Aglai'a, Doris, Pū'lĕs,* and *Virginia;* 1858, *Nemau'sa, Euro'pa, Calyp'so, Alexandra,* and *Pando'ra;* 1859, *Mnemos'ynĕ;* 1860, *Concordia, Dan'aĕ, Olympia, Era'to,* and *Echo;* 1861, *Ausonia, Angeli'na, Cȳb'elĕ, Ma'ia, Asia, Hesperia, Leto, Panope'a, Feronia,* and *Ni'obĕ;* 1862, *Clȳt'iĕ, Galate'a, Euryd'icĕ, Fre'ia,* and *Frig'ga;* 1863, *Diana* and *Euryn'omĕ;* 1864, *Sappho, Terpsich'orĕ,* and *Alcmĕ'nĕ;* 1865, *Beā'trix, Clī'o,* and *Io;* 1866, *Sem'elĕ, Sylvia, This'bĕ,* ⑥⑨, *Antī'opĕ,* ⑨ *;* 1867, ⑨②, ⑨③, ⑨④, ⑨⑤; 1868, ⑨⑥, ⑨⑦, ⑨⑧, ⑨⑨, ⑩⑩, ⑩①, ⑩②, ⑩③, ⑩④, ⑩⑤, ⑩⑥. The largest number discovered in any single year is eleven (in 1869); and in the three years, '57, '61, and '69, no less than thirty were discovered.

d. **Names of the Discoverers.**—*Dr. Luther,* at the observatory of Bilk, near Dusseldorf, has discovered no less than 16, and is at the head of planet discoverers; *Mr. Herman Goldschmidt,* an amateur astronomer of Paris, has discovered 14; *Mr. Hind*, a distinguished English astronomer, 10; *De Gasparis*, at Naples, 9; *M. Chacornac*, at Marseilles and Paris, 6; *Mr. Pogson*, an English astronomer, 6 (3 at Oxford, and 3 at Madras); *Dr. C. H. F. Peters*, at Clinton, N. Y., 8; *M. Tempel*, at Marseilles, 6; *Mr. Ferguson*, at Washington, 3; *Mr. Watson*, at Ann Arbor, Michigan, 9; *Mr. Tuttle*, at Cambridge, Mass., 2; several other observers, 1 or 2 each. Twenty-three of these planets have been discovered in this country. Instead of the names above given, the minor planets are now generally distinguished by *numerals* according to the order of their discovery. Several of these bodies were discovered by two or more observers independently.

295. The AVERAGE DISTANCE of these planets from the sun is about 260 millions of miles. That of the nearest, *Flora*, is about 201 millions; that of the most distant, *Sylvia*, is nearly 320 millions. The entire width of the zone in which they revolve is, however, about 190 millions of miles.

296. The INCLINATION OF THEIR ORBITS is very diverse; more than one-third of the whole have a greater inclination than 8°, and consequently extend beyond the zodiac. The greatest is that of *Pallas*, amounting to 34° 42'; the least, that of *Massilia*, which is only 41'.

297. The ECCENTRICITY of their orbits is equally variable; the most eccentric being that of *Polyhymnia*, which is .337, or more than one-third; the least eccentric is that of *Europa*, which is only .004, or $\frac{1}{250}$.

a. These orbits are not *concentric;* but if represented on a plane surface, would appear to cross each other, so as to give the idea of constant and inevitable collisions. " If," says D'Arrest, of Copenhagen, " these orbits were figured under the form of material rings, these rings would be found so entangled, that it would be possible, by means of one among them taken at a hazard, to lift up all the rest." The orbits do not, however, actually intersect each other, because they are situated in different planes; but some of them approach within very short

distances of each other. The orbit of Fortuna, for example, approaches
the orbit of Metis within less than the moon's distance from the earth.
This is also true of the orbits of Astræa and Massilia, and those of
Lutetia and Juno.

298. The LARGEST of the minor planets is *Pallas,* the
diameter of which is variously estimated at from 300 to 700
miles.   These bodies are generally so small that it is quite
impossible to measure their apparent diameters, or to say
which is the smallest.   The brightest of these planets is
Vesta ; the faintest, *Atalanta.* Vesta, Ceres, and Pallas
have been seen with the naked eye, having the appearance
of very small stars.

299. The SIDEREAL PERIOD of *Flora* is $3\frac{1}{4}$ years; that
of *Sylvia* is about $6\frac{1}{2}$ years.   The average period of the
whole is about $4\frac{1}{2}$ years.

a. **Origin of the Minor Planets.**—The theory of Olbers has
already been alluded to ; it supposes that these little planets are the
fragments of a much larger one, which by an extraordinary catastro-
phe was, in remote antiquity, shivered to pieces.   Prof. Alexander
has endeavored to compute the size and form of this planet   He supposes that it was not of the form of a globe, but shaped like a lens
or wafer, the equatorial and polar diameters being respectively, 70,000
miles and 8 miles ; that the time of its rotation was about $3\frac{1}{2}$ days ;
and that it burst in consequence of its great velocity, as grindstones
and fly-wheels sometimes do.   This theory of an exploded planet has
not been generally accepted, since it is highly improbable, and sup-
ported by no analogous facts.

b. **Nebular Hypothesis.**—This was invented by Laplace to account
for the formation of the solar system by the operation of ordinary
physical laws.   He conceived that the matter of which the various
bodies belonging to this system are composed, originally had an enor-
mously high temperature and existed in the condition of gas or vapor,
filling a vast space ; that as this mass cooled, and, of course, unequally,
currents were formed within it, which, tending to different points or

QUESTIONS.—298. Which is the largest of the planets ? The brightest ? The faintest ?
299. Average sidereal period ? Longest ? Shortest ? *a.* Origin of the minor planets ?
Asteroid planet ? *b.* Nebular hypothesis ?

centres, gave it finally a slow rotation ; that this increased by degrees, until the centrifugal force exceeded the attraction of the central mass, and a zone or ring became detached, of a lower temperature, but still vaporous or liquid ; and that thus successive rings were formed, which breaking up as they rotated, the parts finally came together and formed spheroidal masses revolving around the original mass. If these rings condensed without breaking up they would continue to revolve as rings, like those of Saturn ; if, on the other hand, they broke up into small parts, none sufficiently large to attract all the others, they would condense into fragments and continue to revolve as small planets, like the asteroids. The larger planet masses, being still in a vaporous condition, would, as they cooled and condensed, throw off rings like the original mass , and in this manner either satellites or rings would be formed. The residue of the original nebulous mass he conceived to be the sun.

Such is a brief outline of this celebrated and most ingenious hypothesis,—an hypothesis which every subsequent discovery has seemed to harmonize with and confirm. Whatever theory be adopted to account for the development of the solar system and the existence of this zone of small planets, it must not be forgotten that the infinite power and intelligence of the Great Creator could alone have brought them into being. The only question is, in what way did He exert this power, and in what manner did He ordain that all these wonderful orbs should come into existence as witnesses of His omnipotence and benevolent design.

c. **Decrease in Brightness of the Successive Groups.**—The brightest of the minor planets seem to have been discovered, for each successive group is less conspicuous than those preceding it. The first ten resemble stars of the eighth magnitude [the brightest stars are of the first] ; the last ten are but little brighter than stars of the twelfth magnitude. It is not anticipated, therefore, that others will hereafter be detected with the readiness and frequency which have marked the discoveries of the last ten years. The labor required in the discovery of these little bodies is almost inconceivable. The most successful discoverers have attained the object of their efforts only after mapping down every minute star in certain zones of the heavens ; and to do this required a patient and toilsome watching during every clear night for many months.

CHAPTER XV.

300. The PLANETS, while revolving around the sun, constantly disturb each other's motions, and thus give rise to numerous irregularities, similar to those which take place in the revolution of the moon around the earth.

301. These irregularities are called *inequalities* or *perturbations*. They are either *periodic* or *secular*, the former requiring short, the latter very long periods of time for their completion.

a. **Problem of the Three Bodies.**—To compute the exact place of a planet at any time requires that all the inequalities due to the disturbing action of other planets should be taken into account ; and to do this has tasked to the utmost the highest powers of the human intellect. The problem is, however, simplified by the fact that, as the sun's attraction is so much greater than that of the other bodies, the place of the planet can be found by first supposing that it revolves in an exact elliptical orbit, and then calculating the amount of disturbance due to each other planet in succession ; the aggregate of the results thus obtained giving the proper correction to be applied in order to ascertain the true place. This has been called the *The Problem of the Three Bodies*, because it involves the investigation of the motion of one body revolving around another, and continually disturbed by the attraction of a third. To determine, therefore, all the inequalities to which any planet is subject, it is necessary to solve this problem separately for every other planet by which it may be disturbed. Its complete solution surpasses the powers of the most skillful mathematician.

---

302. The ELEMENTS OF A PLANET'S ORBIT are the facts
which it is necessary to know in order to determine the precise situation of the planet at any instant. They are—1.
The position of the line of nodes ; 2. *The inclination of the
orbit to the plane of the ecliptic ;* 3. *The place of the perihelion ;* 4. *The eccentricity ;* 5. *The major axis.*

a. Elements 1, 2, and 3 determine the *position* of the orbit; 4, its
figure ; and 5, its size. In order to find the place of the planet, it is
necessary also to know the *periodic time,* and the *place of the planet* at
any particular epoch.

b. **Heliocentric and Geocentric Place.**—The true position of a
planet is that in which it would appear to be situated if viewed from
the sun, that is, its *heliocentric* place ; hence, one important point in
ascertaining a planet's true position is to deduce its heliocentric place
from its *geocentric place,* or situation as seen from the earth.

303. The only INVARIABLE ELEMENT is the length of the
major axis ; every other, in the case of each planet, undergoes certain small changes, such as those which have been
described in the orbits and motions of the earth and moon.

a. Thus the *inclinations* of the orbits of Mercury, Venus, and
Uranus are increasing ; those of Mars, Jupiter, and Saturn are diminishing ; the greatest variation being that of Jupiter, which is 23″ in a
century. A similar variation occurs in the *positions* of the *nodes* and
perihelion, and in the amount of *eccentricity.* In the case of the earth,
as has been stated (Art. 125, *e*), the latter is diminishing ; and this is
also true of Venus, Saturn, and Uranus ; while that of Mercury, Mars,
and Jupiter is increasing. The greatest variation is that of Saturn,
which is about .00031 of its mean distance in a century. This is relatively about 7½ times as great as that of the earth, and amounts
absolutely to about 2,700 miles a year ; while the absolute annual
variation of the earth's eccentricity is only 36½ miles. All these
changes are confined within certain very narrow limits, after reaching
which they occur in an opposite direction.

QUESTIONS.—302. What are the elements of a planet's orbit? *a.* What is determined
by them? What else must be known to determine a planet's place? *b.* What is meant
by the heliocentric and geocentric places of a planet? 203. Which element is invariable?
a. What examples are given of variable elements?

304. The MOTIONS OF THE PLANETS are retarded or accelerated by their mutual attractions, according to their positions with respect to each other and to the sun ; but as action and reaction are equal and in opposite directions, whenever one is accelerated the other which acts upon it must be retarded.

Thus, in Fig. 93, page 151, the planet at M must have its motion accelerated by that of the earth at E, while the latter must be retarded ; but the acceleration of M is greater than the retardation of E, because the disturbing force at M acts more nearly in the direction of the planet's motion. After conjunction this is reversed ; the motion of the earth being accelerated and that of the planet retarded.

a. If the planets' orbits were exactly circular, the amount of acceleration in one part of the orbit would be counterbalanced by the retardation in the other, and the irequalities would, in a synodic period, cancel each other : but as the orbits are elliptical, the successive conjunctions must occur at different parts of the orbits, where the planets are at different distances from each other; so that the inequalities must increase while the conjunctions occur in one part of the orbit, and diminish while they take place in the other. If the conjunctions always occurred in the same part of the orbit, the inequalities would constantly accumulate, and the system would be destroyed. This is nearly the case with Jupiter and Saturn.

b. **Great Inequality of Jupiter and Saturn.**—The periodic times of Jupiter and Saturn are respectively 4,332 days and 10,759 days ; and hence, 5 of the former are *nearly* equal to 2 of the latter ; so that, in 5 revolutions of Jupiter, or about 59 of our years, the conjunctions take place at nearly the same points of their orbits. The synodic period of these two planets is 19.86 years : and during the 17th and 18th centuries the conjunctions constantly occurred almost at their points of nearest approach to each other, so that Jupiter's period appeared to be shortened and Saturn's lengthened, greatly to the perplexity of astronomers, till Laplace demonstrated the cause. Similar coincidences exist in the periods of Venus and the earth, but the disturbance accumulates only for a short period. It will be obvious, therefore, that the stability of the system, since the orbits are not *circular*, depends on the periods' being *incommensurable*.

305. Since the attraction of gravitation is reciprocal, the sun is attracted by the planets, and each primary planet is attracted by its satellites; and, therefore, instead of revolving one around the other *as a centre*, they in fact revolve around their common *centre of gravity*.

a. By the *centre of gravity* of two or more bodies connected together in any way, is meant the point around which they all balance each other. The centre of gravity of the solar system moves in a small and very irregular orbit, since it results from the joint action of all the planets. Its distance from the centre of the sun can never be equal to the diameter of the latter; and within this limit the centre of the sun must revolve around it.

306. MASSES OF THE PLANETS.—The amount of attraction exerted by one body upon another is an exact measure of its *mass*. The masses of the planets that are attended by satellites are found by comparing the attraction of the sun upon the planets, with the attraction which they exert themselves upon their satellites. The masses of the planets not attended by satellites are found by ascertaining the amount of disturbance which they occasion in the motions of bodies in their vicinity.

a. Comparative Masses of the Sun and Planets.—To determine these it will be most convenient to resort to simple algebraic representation. Let M be the mass of the sun, and m that of the earth ; F and f, their respective forces of attraction, P and p, their periodic times, and D and d, their distances. Then, according to the law of gravitation, the ratio of the attractions is equal to the direct ratio of the masses multiplied by the inverse ratio of the squares of the distances.

That is, $\dfrac{F}{f} = \dfrac{M}{m} \times \dfrac{d^2}{D^2}$; hence, $\left(\text{dividing by } \dfrac{d^2}{D^2} \right)$ we have $\dfrac{M}{m} = \dfrac{F}{f} \times$

QUESTIONS.—305. Do the planets revolve around the sun *as a centre?* *a.* What is meant by the centre of gravity ? What is the shape and magnitude of the sun's orbit, and the orbit of the centre of gravity ? 306. What is the general method of determining the masses of the planets ? *a.* How to find the comparative masses of the sun and planets ? What calculation is made for the sun and earth ? The earth and Saturn ?

$\dfrac{D^3}{d^3}$. But it can be shown by simple geometry that the forces are directly as the distances and inversely as the squares of the periodic times. That is, $\dfrac{F}{f} = \dfrac{D}{d} \times \dfrac{p^2}{P^2}$. Therefore by substitution, $\dfrac{M}{m} = \dfrac{D^3}{d^3}$ $\times \dfrac{p^2}{P^2}$; that is, *the ratio of the masses is equal to the direct ratio of the cubes of the distances multiplied by the inverse ratio of the squares of the periodic times.* Hence the mass of the sun (that of the earth being one) is $\left(\dfrac{91,430,000}{238,800}\right)^3 \times \left(\dfrac{27.3}{365.25}\right)^2 = 315,000$ (very nearly). In this calculation, we take no account of the attraction of the earth upon the sun or of the moon upon the earth; but this is so small that it would not affect the result materially.

The above formula is applicable to the case of any planet that is attended by satellites. Thus, the masses of the earth and Saturn may be compared by the periodic times and distances of the moon and any of the satellites of Saturn. The distance of Dione is 245,846 miles, and its periodic time about 66 hours; hence the cube of the ratio of the distance of this satellite and that of the moon multiplied by the square of the ratio of their periodic times, or $\left(\frac{245,846}{238,800}\right)^3 \times \left(\frac{66}{655}\right)^2$, will give the mass of Saturn, the earth being 1 By performing the work the result will be found to be 89 +, which is very nearly correct.

The mass of the sun as compared with the earth can also be found by finding the force of gravity at the surface of the earth and comparing it with the force of the sun upon the earth, as determined by the distance and orbital velocity of the latter.

b. From the third of the above formulæ it is obvious that $\dfrac{D^3}{d^3} = \dfrac{M}{m} \times \dfrac{P^2}{p^2}$ and this is evidently applicable to planets revolving around the same central body. But in that case, the mass being the same, $\dfrac{M}{m}$ becomes equal to 1 ; and, therefore, $\dfrac{D^3}{d^3} = \dfrac{P^2}{p^2}$; that is, *the squares of the periodic times are in proportion to the cubes of the mean distances ;* which is Kepler's great law.

QUESTION.—*b.* What demonstration of Kepler's third law is given?

CHAPTER XVI.

COMETS.

307. COMETS are bodies of a nebulous or cloudy appearance that revolve around the sun in very eccentric or irregular orbits, and are generally accompanied by a long and luminous train, called the *tail.*

308. They generally consist of three parts; the *nucleus,* or bright and apparently solid part in the centre; the *coma,* or nebulous substance which envelops it; and the tail, which extends on the side from the sun.

a. The name *comet* is derived from this nebulous appearance which the ancients fancifully likened to hair [in the Greek, *comē*], and hence called these bodies *comētæ,* or *hairy bodies.* When the luminous train precedes the comet, it is sometimes called the *beard.*

b. The appearance of comets is not uniform, the same comet changing very much at different times. Some comets have no nucleus, others, no tails; while still others have several tails.

c. These bodies when at a long distance from the earth and sun are distinguished from planets by the size and position of their orbits, and the direction of their motions. Uranus, it will be remembered, was for some time thought to be a comet, and was recognized as a planetary body only after its orbit had been proved to be almost circular, and nearly in the plane of the ecliptic.

309. Comets either revolve around the sun in elliptic orbits, or move in curve lines called by mathematicians *parabolas* and *hyperbolas.* Elliptic comets may be considered as

---

QUESTIONS.—307. What are comets? 308. Of what parts do they consist? *a.* Origin of the name *b.* Is the appearance of a comet uniform? *c.* How distinguished from planets? 309. In what kind of orbits do they revolve?

210   COMETS.

belonging to the solar system; the others, only as visitants
of it, since they come from distant regions of space, move
around one side of the sun, and then pass swiftly away in
paths that never return into themselves, but are constantly
divergent.

Fig. 110.

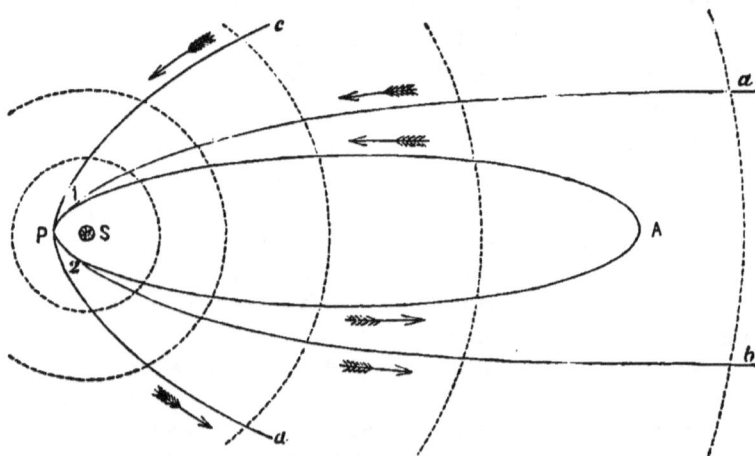

ORBITS OF COMETS.

a. These paths are curve lines of peculiar properties; they are called
" conic sections," because they may be formed by cutting a cone in
various ways. Thus, if a cone be cut by a plane parallel to its base
the curve formed will be a *circle;* if both sides of the cone be cut
obliquely by a plane, the curve will be an ellipse; both of these curves
are continuous lines, returning into themselves. But if the cone be
cut by a plane *parallel to either side* and intersecting the base, the curve
formed will be a *parabola;* and if a plane be passed through the cone
so as to intersect the base at an angle greater than that of the plane
of the parabola, the resulting curve will be a *hyperbola.* The parabola
and hyperbola are not continuous but divergent curves; hence they
do not return into themselves. The parabola is like an ellipse with
only one focus, or an eccentricity infinitely great; and when only a

portion of it is given, it is very difficult to distinguish it from an ellipse. The hyperbola is more easily distinguished, because its arms or branches are more divergent.

In Fig. 110 these three kinds of paths are represented; A and P being the aphelion and perihelion of an elliptic orbit; *a* P *b*, the two branches of a parabolic path ; and *c* P *d*, those of a hyperbolic path. The greater divergency of the last will be obvious; also, that the elliptic and parabolic curves coincide from 1 to 2, so as to be entirely undistinguishable. The motion indicated by the arrows is direct.

310. The ELEMENTS of a comet's orbit are, 1. The *longitude of the perihelion ;* 2. The *longitude of the ascending node ;* 3. The *inclination to the plane of the ecliptic ;* 4. The *eccentricity ;* 5. The *direction of the motion ;* 6. The *perihelion distance from the sun.*

311. The elements of more than 240 cometary orbits have been computed ; and of these only 19 are known to be elliptic, and 5 hyperbolic. The remainder are either parabolic, or elliptic of very great eccentricity.

a. Besides the 19 elliptic comets mentioned, there are 37 that are believed to be elliptic although they have not been proved to be so ; and 11 others more doubtful. There are also 10 doubtful hyperbolic comets ; leaving, out of 242 comets whose elements have been computed, 160 with parabolic orbits, or orbits having an eccentricity too great to be ascertained with accuracy.

312. The ELLIPTIC COMETS are divided into two classes ; those of short periods and those of long periods. The former are seven in number, and have all reappeared several times, their identity being satisfactorily established by an entire correspondence of their elements. The most noted ˙ of these is the comet of Encke, the period of which is about $3\frac{1}{3}$ years, eighteen returns of it having been recorded.

a. The others are *De Vico's,* the period of which is $5\frac{1}{2}$ years; *Winnecke's,* $5\frac{1}{2}$ years , *Brorsen's,* $5\frac{3}{4}$ years; *Biela's,* $6\frac{3}{4}$ years ; *D'Arrest's,*

6⅔ years; *Faye's*, 7½ years. These comets are named after the distinguished astronomers who first discovered them, or determined their periods and predicted their returns. Several others are thought to be comets of short periods.

b. These comets have comparatively small orbits, the mean distance of each being less than that of Jupiter, and all revolving within the orbit of Saturn. The inclination of the orbits is comparatively small, the average being about 12½°. The greatest is 31°, and the least 3°. They all revolve from west to east. They are not conspicuous objects, but have been generally visible only with the aid of a telescope.

313. With the exception of a few comets, the periods of which have been computed to be about 75 years, all the remaining elliptic comets are thought to be of very long periods, some more than 100,000 years.

a. The comet of 1744 is estimated to require nearly 123,000 years to complete one revolution ; that of 1844, 102,000 years; and the great comet of 1680, about 9,000 years. The period of a comet can not, however, be ascertained with precision during one appearance, since only a very small part of its orbit is described during the short time it remains visible. There is, consequently, considerable uncertainty in these determinations. To the great comet of 1811, the two periods of 2,301 and 3,065 years have been assigned.

314. Of all the comets whose orbits have been ascertained, about one-half are direct, that is, revolve from west to east; the remainder are retrograde. Their inclinations are very diverse, some revolving within the zodiac, others at right angles with the ecliptic.

a. There is a decided tendency in the periodic comets to revolve in orbits but little inclined to the ecliptic, while the greatest number of comets are found moving in or near a plane inclined 50° to the ecliptic. Most of the elliptic and hyperbolic comets are direct; of the parabolic, retrograde.

b. About three-fourths of all the comets have their perihelia within the orbit of the earth ; and nearly all the others, within the orbit of the nearest asteroid. Only one is situated more than 400,000,000 miles from the sun. Some comets, on the other hand, come into close proximity to the sun. The great comet of 1680 approached within 600,000 miles of it; and that of 1843 was less than 75,000 miles. The *aphelion* distances of some of these comets are inconceivably great. The comet of 1811 recedes to a distance from the sun equal to 14 times that of Neptune, or more than 40,000 millions of miles; the greatest known (that of 1844) must be nearly 400,000 millions of miles.

The aphelion distance can be found from the eccentricity and perihelion distance. The latter in the case of the comet of 1844 is about 80,000,000 miles; the eccentricity, .9996 of the semi-axis. Hence 1 — .9996 =.0004 of the semi-axis must be the perihelion distance; and 80,000,000 ÷ .0004 = 200,000,000,000 = semi-axis.

c. The **velocity** of comets as they move through their perihelia is amazingly great. That of 1680 was 880,000 miles an hour; and that of 1843, about 1,260,000 miles an hour, or 350 miles per second. The latter body swept around the sun from one side to the other in about two hours.

315. The NUMBER OF COMETS is supposed to be very great. From the earliest period up to the present time more than 800 have been recorded, of which nearly 300 have had their orbits computed, and of the latter 54 have been identified as returns of previous comets.

a. Since it is only within the last 100 years that optical aid has been made available in searching for comets, it is supposed that the actual number of comets that have come within view, in both hemispheres, is not less than 4,000 or 5,000. M. Arago estimates that the greatest possible number in the solar system can not exceed 350,000.

316. The SIZE of comets, including both envelope and nucleus, very much exceeds that of the largest planet; the nucleus is, however, comparatively small, the diameter of the largest measured being about 8,000 miles (that of 1845).

a. The nucleus of the comet of 1858 (Donati's) was 5,600 miles in diameter; that of 1811, only 428 miles. The *coma* of the latter was found to

---

COMETS.

be 1,125,000 miles; and that of Encke, 281,000 miles. The dimensions of comets, however, vary greatly at different parts of their orbits, contracting as they approach the sun, and expanding as they recede from it. Thus Encke's comet in October, 1838, was more than 250,000 miles in diameter; but in December, contracted to 3,000 miles.

317. The MASSES AND DENSITIES of the comets must be inconceivably small; since, notwithstanding their great magnitudes, they move among the planets and their satellites without in the least, as far as it can be observed, affecting their motions; although they are themselves greatly disturbed by the attractions of the planets.

a. Their densities are, without doubt, many thousand times less than atmospheric air. Stars are seen very clearly through the nebulous coma and train of a comet, notwithstanding that the light has to pass sometimes through millions of miles of the substance.

318. The TAILS of comets are often of immense length, and are generally of a bent or curved form, extending on the side from the sun and nearly in a line with the radius-vector of the orbit. The tail increases in length as the comet approaches the sun, but attains its greatest dimensions a short time after the perihelion passage, and then gradually diminishes.

a. In respect to magnitude, the tails of comets are the most stupendous objects which the discoveries of astronomers have presented to our contemplation. That of the comet of 1680 was more than 100,000,-000 miles in length, while the comet of 1843 presented a train 200,000,000 miles long, which was shot forth from the head of the comet in the incredibly short space of twenty days. The increase of the tail and the decrease of the head of the comet as it approaches the sun, are among the most striking phenomena presented by these bodies

b. The tails of comets are not of uniform breadth, but diverge or spread out as they extend from the head. The middle of the tail usually presents a dark stripe which divides it longitudinally into two parts. This appearance is usually explained by the supposition that

QUESTIONS.—317. Masses and densities? *a.* Why is the density thought to be small? 318. Position and length of tails? *a.* Examples? Change in length at different times? *b.* Appearance of tails? How explained?

the tail is hollow, being a kind of conical shell of vapor; and as we look through a considerable thickness of the vapor, at the edges, it appears brighter there than in the middle where the quantity is comparatively small.

c. The diminution of the size of comets as they approach the sun is probably to some extent only apparent; since their substance must necessarily be vaporized as they approach the sun, and much of it so attenuated as to become invisible. There is no doubt also that a considerable portion is exhausted in the formation of the tail; and that as the comet moves in its orbit it loses by disruption considerable portions which pass away into space.

319. Observations with the polariscope have shown that the tails of comets shine by reflected light; but that the nucleus and coma emit quite a strong radiance of their own.

a. If the head of the comet shone by reflected light alone, its apparent brightness would be inversely proportional to the product of the squares of the distances from the sun and earth; but this is contrary to observation. Donati's comet (that of 1858), according to this rule, should have been 188 times as bright when near its perihelion in October as it was in June; whereas it was actually 6,300 times as bright, its own light having increased in the ratio of 33 to 1.

b. Some astronomers suppose the nucleus to be a solid, partially or wholly converted into vapor by the intense heat of the sun; others, that it is of the same nature as the coma, only more dense. It was the opinion of Sir William Herschel, and is still a very generally accepted one, that the nucleus is surrounded with a transparent atmosphere of vast extent, within which the nebulous envelop floats like clouds in the earth's atmosphere. This nebulous matter appears to be continually driven off by some force emanating from the sun, and thus f rms the luminous train. At their perihelia comets must generally be subjected to a heat far more intense than would be required to melt the hardest substance found on the surface of the earth. Prof. Norton thinks that the tail is formed by two streams, in opposite magnetic or electric states, expelled from opposite points, or poles, of the nucleus, and bent back by the sun's repulsive force until they nearly meet, being separated by only a narrow interval, which appears as the dark stripe noticed in the tail.

REMARKABLE COMETS.

320. COMET OF 1680.—This was the comet that Newton subjected to the calculations by which he showed that these

Fig. 111.

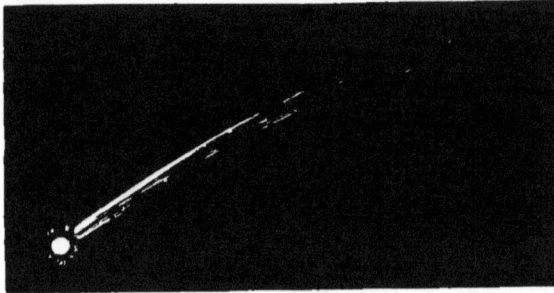

GREAT COMET OF 1680.

bodies revolve in one of the conic sections, and that they are retained in their orbits by the same force that binds the planets to the sun. It was very remarkable for its splendor, and for the extent of its train, which stretched over an arc of 70° in the heavens, and reached the amazing length of 120,000,000 miles. With the exception of the comet of 1843, it approached nearer to the sun than any other known, and moved through its perihelion with a velocity of 880,000 miles an hour.

a. Its perihelion distance is .0062 (the earth's distance being 1), and its eccentricity, according to Encke, is .99998. Now, 91,500,000 × .0062 = 567,300; and 1 — .99998 = .00002. Hence 567,300 ÷ .00002 = 28,365,000,000, which is its semi-axis; and if we multiply this by 2, and subtract the perihelion distance from the product, we shall find the aphelion distance, which is equal to nearly 57,000 millions of miles. The period corresponding to this orbit is 8,814 years. Some ascribe to this comet a much shorter period; and others, a hyperbolic orbit.

321. HALLEY'S COMET. — This comet derives its name from Sir Edmund Halley, a celebrated English astronomer, who calculated its orbit and predicted its return. It appeared in 1682, and Halley noticing a close resemblance

in its elements to those of
1531 and 1607, concluded
that the comets of these
years were different appearances of the same comet, and
ventured to predict its re-
appearance in 1758 or 1759.
This prediction was real-
ized by the return of the
comet in March, 1759; and
it again appeared in 1835.
These different appearances,

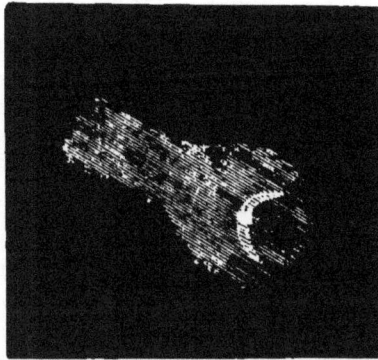

Fig. 112.

HALLEY'S COMET, 1835.

it will be observed, were about 75 years apart ; and others of
an earlier date have also been recognized.

a. **History of the Prediction.**—This celebrated prediction of
Halley may be considered almost the first fruits of Sir Isaac Newton's
demonstration of the laws of planetary motion as contained in his
famous work, the *Principia*, published in 1687.   The comet of 1682
had been an object of interest to both Halley and Newton, and its
path had been calculated by Picard, Flamstead, and others.   It occurred to Halley that this comet might be identical with others previously
recorded ; and fortunately the comet of 1607 had been observed by
Kepler and Longomontanus, and that of 1531, by Apian at Ingolstadt ;
the path in each case being quite accurately determined.   The coincidence which Halley noticed in these paths gave him confidence in the
prediction which he made.   He observed, however, that as the comet
in the interval between 1607 and 1682, passed near Jupiter, its
velocity must have been increased and its period shortened ; so that
the next interval would be 76 years or upward, and the comet would
return at the end of 1758 or the beginning of 1759.   Subsequent
researches gave increased force to this prediction ; for it appeared that
comets had been seen in 1456 and 1378, whose paths seemed to have
been nearly identical with that of the comet of 1682.

b. **The Prediction Realized.**—As the time drew near, the attention
of the scientific world was awakened to the subject ; and it was

resolved to compute more exactly the time of the comet's appearance, by applying all the additional resources of mathematical science that seventy-five years had brought forth. This was a gigantic undertaking, since it was necessary to calculate the distance of each of the two planets, Jupiter and Saturn, from the comet, and the exact amount of their disturbance, separately, for every successive degree, and for two revolutions of the comet, or 150 years. Clairaut and Lalande, two French mathematicians, undertook the work, the latter being assisted in the arithmetical portion of it by Madame Lapaute; and after six months spent in calculations, from morning to night, this enormous sum was worked out, and the day of the comet's return to its perihelion was announced. This was April 11th. It actually passed its perihelion March 13th, or about 22 days previously to the predicted time. Clairaut, however, stated in announcing his prediction that the comet might be accelerated or delayed by the attraction of an undiscovered planet beyond the orbit of Saturn, thus anticipating, in imagination, the discovery of Uranus which Herschel made 22 years afterward. Halley did not live to witness the realization of his prediction, having died in 1742.

c. **The Return in 1835.**—The time of its perihelion passage in 1835 was computed by several mathematicians, the mean of all the results being November 12th. The comet was observed to pass its perihelion on the 16th of that month. It continued visible in the southern hemisphere for several months, and then disappeared, not to be seen again until 1911.

d. The **mean distance** of this comet is a little less than that of Uranus. Its perihelion distance is about 60 millions of miles; its aphelion distance more than 3,200 millions. Its motion is retrograde, and the inclination of its orbit about 18°. History shows that it has regularly returned during a period of more than 18 centuries, its first recorded appearance being in 11 B.C. It seems however, to have been a far more conspicuous object in its ancient visitations than at its more recent returns. In 1066 and 1456, it was an object of immense size and splendor, and created wide-spread alarm.

322. ENCKE'S COMET is remarkable for its short period and frequent returns. Its period and elliptic orbit were deter-

mined by Professor Encke
at its fourth recorded appearance in 1819. Its
last return took place in
1868; the next will occur
in January, 1872. This
comet has generally ap-
peared without any lumi-
nous train; but in 1848,
it had a tail about 1°
long, turned from the

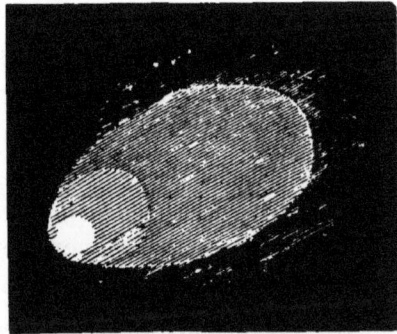

Fig. 113.

ENCKE'S COMET.

sun, and a shorter one directed toward that luminary. In
its latest returns it has been very faint and difficult of
observation.

a. **Mass of Mercury.**—The return of 1838 led to the establishment
of an important fact. In August, 1835, this comet passed very near
Mercury; and Encke showed that, if Laplace's value of Mercury's mass
were correct, the comet's motion would be greatly disturbed; but as
this was found not to be the case, it was obvious that the received
determination of Mercury's mass needed correction. A much lower
value has since been adopted; but astronomers do not entirely agree
as to this element. Encke's value is about $\frac{1}{16}$ that of the earth; but
Leverrier's is a little more than $\frac{1}{11}$. Laplace's had been about $\frac{1}{6}$.

b. **The Resisting Medium.**—A still more interesting discovery has
been evolved from observations of this comet. Professor Encke found
that at each return, the arrival of the comet at its perihelion took
place about 2¼ hours earlier than the most exact calculations predicted,
and that this constant acceleration had amounted since 1786 to about
2¼ days. As this could not be attributed to the disturbing influence of
any unknown body, he conceived that it could be caused only by a
resisting medium filling the interplanetary spaces; since the effect of
such a medium would be to diminish the centrifugal force, and thus
bring the body nearer to the sun : so that its orbit would be contracted and its periodic time made constantly shorter. A very ethereal
fluid would be sufficient to produce this result in the case of a body so
light as a comet; while it would have no appreciable effect on the

QUESTIONS.—*a.* How was the mass of Mercury found? *b.* Resisting medium?

planets on account of their great mass and enormous momentum. A similar acceleration takes place in the case of Faye's comet.

323. LEXELL'S COMET.—This body is particularly noted for the amount of disturbance which it has suffered in passing among the planets. From observations made in 1770, Lexell calculated its period at about $5\frac{1}{2}$ years; and it was a large and bright object, the diameter of its head being about $2\frac{1}{3}°$. It has, however, never been seen since, its orbit having been entirely changed by planetary disturbance.

a. Investigation showed that it really returned in 1776, but was so situated as to be continually hid by the sun's rays; that in 1779, it passed so near Jupiter that its orbit was greatly enlarged, so that it no longer comes near the earth. The fact that it never appeared previous to 1770, is accounted for in a similar way; its orbit having in 1767 been changed by the attraction of Jupiter, from one of large to one of small dimensions. On July 1st, 1770, the distance of this comet from the earth was less than 1,500,000 miles.

Fig. 114.

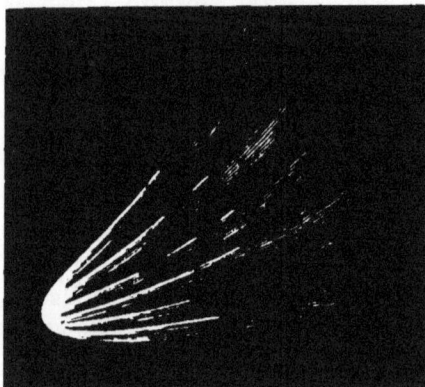

COMET OF 1744.

324. COMET OF 1744.— This was the finest comet of the 18th century, and according to some observers, had six tails spread out in the form of a fan. Euler calculated its elliptic orbit, and assigned to it a period of 122,683 years. Its motion was direct.

325. BIELA'S COMET.— This is one of the elliptic comets of short period; its perihelion lying just within the orbit of the earth, and its aphelion a little beyond that of Jupiter. The orbit of this body

QUESTIONS.—323. Lexell's comet—why noted? *a.* How accounted for? 324. Comet of 1744? 325. Biela's comet?

nearly crosses the actual path of the earth; and in 1832, Olbers calculated that it would come within 20,000 miles of the earth, so that the latter body would be enveloped in its mass. The earth, however, did not reach the node until one month after the comet had passed it.

a. In 1845, this comet became elongated in form and finally separated into two comets, which traveled together for more than three months; their greatest distance apart being about 160,000 miles. The two parts were again seen at the next return of the comet in 1852, but the interval had increased to 1,250,000 miles. It has not been seen since

Fig. 115.

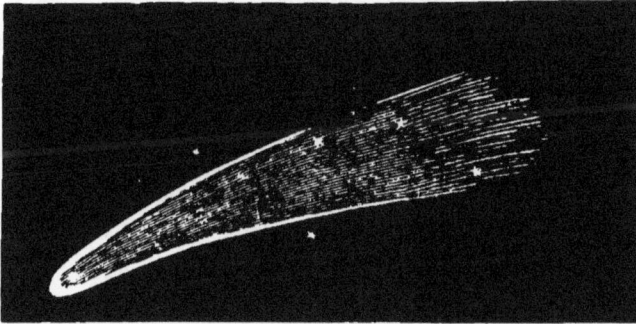

COMET OF 1811.

326. COMET OF 1811.—This comet was very remarkable for its unusual magnitude and splendor. It was attentively observed by Sir William Herschel, who describes it as having a nucleus 428 miles in diameter, which was ruddy in hue, while the nebulous mass surrounding it was of a bluish-green tinge. Its tail was of peculiar form and appearance, extending about 25°, with a breadth of nearly 6°.

a. The investigation of its elements by Argelander is the most complete ever made. He assigns it a period of more than 3,000 years, and estimates its aphelion distance at 40,121 millions of miles.

327. COMET OF 1843.-This comet was also remarkable for its

Fig. 116.

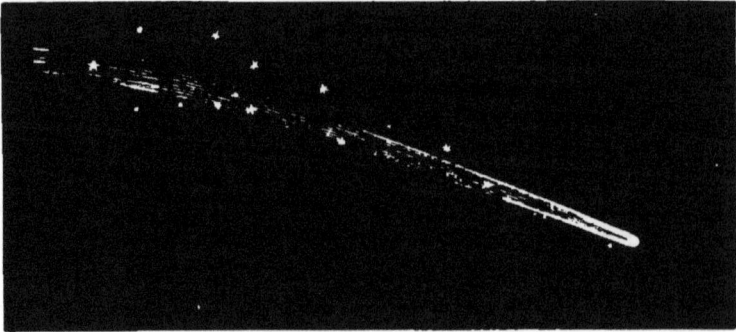

GREAT COMET OF 1843.

extraordinary size and splendor, it being visible in some parts
of the world during the day time. It had a tail 60° long, and
approached within a very short distance of the sun,—about
75,000 miles from its surface. Its period is variously estimated at from 175 to 376 years. Its motion is retrograde.

328. DONATI'S COMET.—This is the great comet of 1858,
named after Donati, by whom it was first seen at Florence.
As it approached its perihelion it attained a very great magnitude and splendor, and was particularly distinguished for
the magnificence of its train. Its period has been estimated
at nearly 1,900 years.

329. RECENT COMETS.—About thirty comets have appeared since that of Donati, the elements of which have
been calculated. The most remarkable were the comet of
1861, described as one of the most magnificent on record,
having a tail 100° long; and that of 1862, which was very
interesting for the peculiar phenomena which it presented
of luminous jets, issuing in a continuous series from its
nucleus.

CHAPTER XVII.

METEORS OR SHOOTING STARS.

330. METEORS* or SHOOTING STARS are small luminous bodies that move rapidly through the atmosphere, followed by trains of light, and quickly vanishing from view. They sometimes appear in numbers so great as to seem like showers of stars.

a. Star-Showers Periodical.—These star-showers are found to occur at certain periods. Every year, about November 14th, there is a larger fall than usual of meteors ; but about every 33 years, it has been noticed, there is a great star-shower. Those which occurred in November, 1866-7, had been predicted from observations of previous events of the kind. Thus a star-shower occurred in November, 1832-3, also in 1799; and there are eighteen recorded observations of the phenomena from 1698 to 902, all corresponding in period to that mentioned above.

b. Great Star-Showers.—The shower of 1799 was awful and sublime beyond conception. It was witnessed by Humboldt and his companion, M Bonpland, at Cumaná, in South America, and is thus described by them :—" Toward the morning of the 13th of November, 1799, we witnessed a most extraordinary scene of shooting meteors. Thousands of *bolides* and falling stars succeeded each other during four hours. Their direction was very regularly from north to south, and from the beginning of the phenomenon there was not a space in the firmament, equal in extent to three diameters of the moon, which was not filled every instant with bolides or falling stars. All the meteors

---

* From the Greek word *meteora*, meaning *things in the air.*

QUESTIONS.—330. What are meteors? *a.* What periods have been observed in their occurrence ? *b.* What instances of great showers?

left luminous traces, or phosphorescent bands behind them, which lasted seven or eight seconds." The same phenomena were witnessed throughout nearly the whole of North and South America, and in some parts of Europe. The most splendid display of shooting stars on record was that of November 13th, 1833, and is especially interesting as having served to point out the periodicity in these phenomena. Over the northern portion of the American continent the spectacle was of the most imposing grandeur; and in many parts of the country the population were terror-stricken at the awfulness of the scene. The ignorant slaves of the southern States supposed that the world was on fire, and filled the air with shrieks of horror and cries for mercy. The shower of 1866 was anticipated with great interest; and in New York and other places arrangements were made to announce the occurrence, during the night of November 14th, by ringing the bells from the watch-towers. The display, however, was not witnessed in this country, but in England was quite brilliant; as many as 8,000 being counted at the Greenwich observatory. Another shower of less extent occurred in November, 1867.

331. METEORIC EPOCHS are particular times of the year at which large displays of shooting stars have been observed to occur at certain intervals. The principal of these are November 13th–14th, and August 6th–11th.

a. Three others have been established with considerable certainty; namely, in January, April, and December, and still others indicated, that are doubtful. There are 56 meteoric days in the year; those in August and November being the richest.

b. **August Meteors.**—Of 315 recorded meteoric displays, 63 seem to have occurred at this epoch. The first eleven, with one exception, were observed in China, between 811 A.D., and 933 A.D., and occurred a few days previous to August 1st. The period of this shower is exactly the same as the sidereal year ; and therefore it occurs about a day later in 71 tropical or civil years. Its maximum period is much longer than that of the November meteors, being estimated at 105 years.

332. METEORS are supposed to be small bodies collected in rings or clusters, and revolving around the sun in eccentric orbits. They appear to resemble comets in their nature and origin, and, like those bodies, sometimes revolve from east to west.

a. **Origin of Meteors.**—The immense velocity of these bodies, which is about equal to twice that of the earth in its orbit, or 36 miles a second, and the great elevation at which they become visible, the average being 60 miles, indicate that they are not of terrestrial, but *cosmical,* origin ; that is, they emanate from the interplanetary regions, and being brought within the sphere of the earth's attraction, precipitate themselves upon its surface. Moving with so great a velocity through the higher regions of the air, they become so intensely heated by friction that they ignite, and are either converted into vapor, or, when very large, explode and descend to the earth's surface as meteoric stones, or *aerolites.** The brilliancy and color of meteors are variable ; some are as bright as Venus or Jupiter. About two-thirds are white ; the remainder yellow, orange, or green.

b. **Number of Meteors.**—The average number of shooting stars seen in a clear, moonless night by a single observer is 8 per hour ; a sufficient number of observers would perceive 30 per hour, which is equivalent to 720 per day, seen by the naked eye at any point of the earth's surface, if the sun, moon, and clouds were absent. But the number visible over the whole earth is about 10,500 times that seen at a single point ; and therefore the average number daily entering the atmosphere, and sufficiently large to be seen by the naked eye, is more than 7½ millions; while at least 50 times as many can be seen through the telescope ; so that about 400 millions must descend to the earth during each day. It becomes therefore an interesting question how much foreign matter may be added to the earth and its atmosphere by these meteoric falls.

333. FIRE BALLS are large meteors that make their appearance at a great height above the earth's surface, moving with immense velocity, and accompanied by luminous

*From the Greek word *aer,* meaning the *air,* and *lithos,* a *stone.*

trains. They generally explode with a loud noise, and sometimes descend to the earth in large masses.

a. No deposit has been known to reach the earth from ordinary shooting stars; probably, because being very small they are dissipated in the air; but scarcely a year passes without the fall of aerolites in some parts of the earth, either singly or in clusters. Some estimate the whole number that fall annually at 700; others, much higher. The most ancient fall of meteoric stones on record is that mentioned by Livy, which occurred on the Alban Hill, near Rome, about the year 654 B.C. There are very many remarkable occurrences of this kind on record, some of the masses being of immense size, and the explosion so violent as to sound like thunder. In 1783 a fire ball of extraordinary magnitude was seen in Scotland, England, and France. It produced a rumbling sound like distant thunder, although its height was 50 miles when it exploded. Its diameter was estimated at about half a mile, and its velocity was as great as that of the earth in its orbit. In 1859, between 9 and 10 o'clock A.M., a meteor of immense size was seen in the eastern part of the United States. Its apparent diameter was nearly equal to that of the sun; and it had a train several degrees in length, plainly visible in the sunshine. Its disappearance on the coast of the Atlantic was followed by several terrific explosions. Some of these meteors have been supposed to pass the earth, moving away into space; others to revolve in an orbit around it, becoming small satellites. A French astronomer assigns to one of the latter a period of revolution of 3 hours and 20 minutes, and a distance from the earth of 5,000 miles.

b. **Composition and Size of Aerolites.**—The materials composing these bodies are always nearly the same, consisting largely of iron, and in no case of any other elementary substances than are found on the earth. Some have been discovered of immense size; one, a mass of iron and nickel, found in Siberia, weighs 1,680 lbs. At Buenos Ayres there is a mass partly buried in the ground $7\frac{1}{2}$ feet in length, and supposed to weigh about sixteen tons. A similar block, weighing about six tons, was discovered a few years ago in Brazil. Many others exist. All these are doubtless of cosmical origin, having been very small

---

QUESTIONS.—*a.* Frequency of the fall of aerolites? Earliest recorded instance? Remarkable instances? Do they all reach the earth? *b.* Composition of aerolites? Their size? Additions to the earth, Venus, and Mercury from this cause? Effect on Mercury's period?

planets revolving around the sun, but brought within the earth's attraction ; and there is no doubt that, before the solar system had reached its present condition, the additions made to the matter of the earth in this way were quite considerable. This is supposed still to be the case with Venus and Mercury, moving as they are through the thicker portions of the great ring which we call the zodiacal light. Now, as Mercury's orbit is very eccentric, it receives at its aphelion a large number of these meteors whose periods are longer than its own ; and this would have the effect to diminish its mean motion and lengthen its period. Such an effect has actually been discovered.

c. **Meteoric Dust, etc.**—There are on record many instances of showers of dark-colored dust, which have fallen from the higher regions of the atmosphere, and which seem from the composition of the dust to be of meteoric origin. These falls are often preceded or attended by a flashing of light as well as by a loud noise, sometimes resembling thunder. In March, 1813, a shower of *red* dust fell in Tuscany, discoloring the snow which then lay on the ground ; and at the same time, a few miles distant, there occurred a shower of aerolites, lasting about two hours, and accompanied by a noise as of the dashing of waves The phenomena of *black* and *red rain* and *snow* are attributed to a similar cause. Since, as has been shown, several millions of meteors pass into the atmosphere during the year, there is no doubt that large quantities of dust, too fine to be visible, descend to the earth's surface. Some of this dust has been detected upon the tops of mountains in soil which had never been previously disturbed by man. Partial obscurations of the sun's light, occasioning what are recorded as *dark days*, and the passage of large black masses across the sun's disc, too rapid to be spots, are probably meteoric phenomena.

334. The NOVEMBER METEORS are supposed to revolve around the sun in an orbit of considerable eccentricity, inclined to the plane of the ecliptic in an angle of $17\frac{1}{2}°$, and extending at its aphelion somewhat beyond the orbit of Uranus, its perihelion being very nearly at that of the earth. They move in a ring of unequal width and density, the

thickest part crossing the earth's orbit every 33 years, and requiring nearly two years to complete the passage.

a. The elements of this orbit correspond almost precisely with those of the comet which made its appearance in January, 1866; so that it seems probable that the comet is a very large meteor of the November stream. The elements of the orbit of the August meteors have been found, in a similar manner, to coincide with those of the third comet of 1862; showing that the comet and these meteors belong to the same ring. This seems also to be true of the first comet of 1861 and the April meteors.

b. The point from which the November meteors seem to radiate is in the constellation Leo; because, as the earth at that time of the year is moving toward that point, they appear to rush from it. Their velocity appears to be double that of the earth, although only equal to it; because they move in an opposite direction and almost in the same plane. When the earth plunges into the meteoric stream a great star-shower occurs.

c. **Physical Origin.**—Meteors are supposed by some to be small fragments of nebulous matter detached in vast numbers from the larger masses which are seen in the regions of the stars, or from that of which the solar system was originally formed, their origin being precisely the same as that of the comets, which indeed may be considered as, in reality, only meteors of vast size. It is also probable that, like Biela's comet, others have been divided and subdivided so as finally to be separated into small fragments moving in the orbit of the original comet, and thus constituting a meteoric ring or stream.

d. The following general conclusions with regard to meteors in the solar system have been suggested: 1. Biela's comet in 1845 passed very near, if not through, the November stream, and was probably divided in this way; 2. The rings of Saturn are dense meteoric streams, the principal or permanent division being due to the disturbing influence of the satellites; 3. The asteroids are a stream or ring of meteors, the largest being the minor planets which have been discovered; 4. The meteoric masses encountered by Encke's comet may account for the shortening of the period of the latter without the hypothesis of a resisting medium.

QUESTIONS.—*a.* Resemblance to comets? *b.* Radiant point of November meteors? *c.* Physical origin of the meteors? *d.* Generalizations with regard to meteors in the solar system?

CHAPTER XVIII.

335. The STARS are luminous bodies like the sun, out situated at so vast a distance from the earth that they seem like brilliant points, and always in nearly the same positions with respect to each other.

a. The **scintillation or twinkling** of the stars is due to the inequalities in density, moisture, etc., of the different strata of the atmosphere through which the rays of light pass. In tropical regions, where the atmospheric strata are more homogeneous, this scintillation is rarely observed; so that, as remarked by Humboldt, "the celestial vault of these countries has a peculiarly calm and soft character."

b. **Parallax of the Stars.**—The usual method of finding the parallax of a body by viewing it at different parts of the earth's surface is entirely useless in the case of the stars, as the displacement thus occasioned in the positions of any of them is utterly inappreciable; the radius of the earth at a distance so immense being practically but a mathematical point. If, however, we view the same star at intervals of six months, our stations of observation will be about 180 millions of miles apart; and the amount of displacement thus occasioned, when reduced to the centre of the orbit, is the stellar parallax, called sometimes the *annual parallax.*

336. The ANNUAL PARALLAX is the change which would take place in the position of a star if it could be viewed from the centre of the orbit, instead of the orbit itself.

a. In other words, it is the angle subtended by the semi-axis of the

---

earth's orbit at the distance of the star. The greatest parallax yet discovered in the case of any star is somewhat less than 1″ (0.9187″), so that the earth's orbit itself is but little more than a mere point at the nearest star. To determine this small parallax exactly is probably the most difficult problem in practical astronomy.

b. **Distance Calculated.**—The sine of 0.9187″ is about .000004464, which is the ratio of the semi-axis of the earth's orbit to the distance of the star. Hence the distance of the star must be 224,000 times the semi-axis of the earth's orbit, or 91½ millions of miles ; and 91,500,000 × 224,000 = 20,496,000,000,000 miles, or nearly 20½ trillions of miles. Light, moving with a velocity of 184,000 miles a second, requires more than 3½ years to pass across this enormous interval,—an interval more than 7,000 times the distance of Neptune from the sun. However large the stars may be, therefore, their attraction upon the solar system must be altogether too feeble to disturb the motions of its component bodies in the least. The parallax of *twelve* stars has been determined with considerable precision, the smallest being 0.046″, or about one-twentieth that mentioned above ; this star must therefore be about 410 trillions of miles from us,—a distance which light would not traverse in less than 70.3 years.

337. MAGNITUDES.—The stars are divided into classes according to their apparent brightness, the brightest being distinguished as stars of the first magnitude, the next of the second, and so on. Stars of the first six magnitudes are visible to the naked eye ; but the telescope reveals the existence of others so feeble in light as to be classed as of the seventeenth magnitude.

a. This classification is based exclusively on appearance, and indicates nothing as to the real magnitudes of the bodies in question. Sir John Herschel gives the following comparative estimate of the amount of light emitted by stars of the first six magnitudes :

6th magnitude	=	1	3d magnitude	=	12
5th "	=	2	2d "	=	25
4th "	=	6	1st "	=	100

This is not uniformly the relative brightness of stars thus denominated; Sirius, the brightest star in the heavens, being 324 times as brilliant as an average star of the 6th magnitude.

338. The WHOLE NUMBER of stars visible to the naked eye in the northern hemisphere is about 2,400; in both hemispheres, more than 4,500.

a. These are distributed by Argelander according to their magnitudes as follows: 1st magnitude, 9 ; 2d, 34 , 3rd, 96 ; 4th, 214 ; 5th, 550 ; 6th, 1439 ; total in northern hemisphere, 2,342. If the southern hemisphere is equally rich in stars, the whole number must be 4,684 ; some estimate it at 6,000 or 7,000. The stars are probably less bright in proportion as their distance is greater; and hence the number increases as we descend to the lower magnitudes. Argelander's estimate for the 9th magnitude is 142,000. Viewed through the telescope, the stars can be counted by millions.

THE CONSTELLATIONS.

339. To facilitate the naming and location of the stars, the heavens are divided into particular spaces, represented on the globe or map as occupied by the figures of animals or other objects. These spaces and the groups of stars which they contain are called *constellations,* or *asterisms.*

a. Thus there are the constellations *Aries,* the Ram ; *Leo,* the Lion ; *Gemini,* the Twins, etc. The general position of a star, according to this system, is defined by stating in what part of the figure it is situated ; as, the *eye of the Bull, the heart of the Lion,* etc. Its exact position is, of course, only to be defined by its right ascension and declination, or longitude and latitude. This system of grouping the stars into constellations is supposed to be very ancient. Ptolemy counted only forty-eight constellations ; but, since his time, the number has been augmented to 109.

340. The stars belonging to each constellation are distinguished by particular names ; as *Sirius, Regulus, Arcturus,* etc., and by letters or numerals.

a. Only the most conspicuous stars have particular names ; the most usual mode of designation being the use of the letters of the Greek alphabet, *alpha* (*a*) being given to the brightest star, *beta* (*β*) to the next, and so on. When the twenty-four letters of this alphabet are exhausted, the Roman letters are used, and subsequently the Arabic numerals, the latter being applied according to the positions of the stars in the constellation, the most eastern being designated 1, which is thus the first star to cross the meridian.

b. **The Greek Alphabet.**—The following are the letters of the Greek alphabet, with their names. It will be convenient for the student to become familiar with them, as they are very frequently employed.

a	Alpha	η	Eta	ν	Nu	τ	Tau
β	Beta	θ	Theta	ξ	Xi	υ	Upsilon
γ	Gamma	ι	Iota	o	Omicron	φ	Phi
δ	Delta	κ	Kappa	π	Pi	χ	Chi
ε	Epsilon	λ	Lambda	ρ	Rho	ψ	Psi
ζ	Zeta	μ	Mu	σ	Sigma	ω	Omega

341. The constellations are distinguished as Northern, Zodiacal, and Southern, according to their positions in the heavens with respect to the ecliptic. The zodiacal constellations have the same names as the signs, but are situated about 28° to the east of them, so that *Aries*, although the first sign of the ecliptic, is the second constellation of the zodiac (Art. 105, *b*).

342. The WHOLE NUMBER of constellations is 109; but many of them are not generally acknowledged or much used by astronomers at the present time.

a. The following catalogue contains the names of the principal constellations, with their right ascension and declination, the number of stars of the first five magnitudes contained in each, and the name of the astronomer by whom they were first enumerated or invented :

NOTE.—The right ascension and declination of the central points of the constellations are given.

THE NORTHERN CONSTELLATIONS.

No.	NAME.	MEANING.	BY WHOM ENUMERATED OR INVENTED.	STARS. No. OF	R. A.	DEC.
1	ANDROMEDA,	The *Chained Princess*,	Ptolemy, 150 A.D.	18	15°	85°
2	AQUILA,	The *Eagle*,	Ptolemy.	83	292½°	10°
3	AURIGA,	The *Charioteer*,	Ptolemy.	35	90°	42°
4	BOOTES.	The *Bear Driver*,	Ptolemy.	35	219°	31.°
5	CAMELOPARDALUS,	The *Giraffe*,	Hevelius, 1690.	36	86°	68°
6	CANES VENATICI,	The *Hunting Dogs*,	Hevelius, 1660.	15	195°	40°
7	CASSIOPEIA,	The *Queen in her Chair*,	Ptolemy.	46	17½°	60°
8	CEPHEUS,	The *King*,	Ptolemy.	44	325°	65°
9	CLYPEUS SOBIESKII,	*Sobieski's Shield*,	Hevelius, 1690.	4	272½°	15° S
10	COMA BERENICES,	*Berenice's Hair*,	Tycho Brahe, 1618.	20	190°	25°
11	CORONA BOREALIS,	The *Northern Crown*,	Ptolemy.	19	235°	30°
12	CYGNUS,	The *Swan*,	Ptolemy.	67	315°	42°
13	DELPHINUS,	The *Dolphin*,	Ptolemy.	10	310°	15°
14	DRACO,	The *Dragon*,	Ptolemy.	80	260°	66°
15	EQUULEUS,	The *Little Horse, or Horse's Head.*	Ptolemy.	5	315°	6°
16	HERCULES,	*Hercules*,	Ptolemy.	65	251°	27°
17	LACERTA,	The *Lizard*,	Hevelius, 1690.	13	335°	44°
18	LEO MINOR,	The *Lesser Lion*,	Hevelius, 1690.	15	161°	36°
19	LYNX,	The *Lynx*,	Hevelius, 1690.	28	120°	50°
20	LYRA,	The *Harp*,	Ptolemy.	18	280°	35°
21	PEGASUS,	The *Winged Horse*,	Ptolemy.	48	336°	15°
22	PERSEUS ET CAPUT MEDUSÆ,	*Perseus & Medusa's Head*,	Ptolemy.	40	52½°	47°
23	SAGITTA,	The *Arrow*,	Ptolemy.	5	295°	18°
24	SERPENS,	The *Serpent*,	Ptolemy.	23	275°	10°
25	TAURUS PONIATOWSKII,	*Poniatowski's Bull*,	Poczobut, 1777	6	267½°	5°
26	TRIANGULUM,	The *Triangle*,	Ptolemy.	5	80°	82°
27	URSA MAJOR,	The *Great Bear*,	Ptolemy.	63	160°	58°
28	URSA MINOR,	The *Lesser Bear*,	Ptolemy.	23	225°	78°
29	VULPECULA ET ANSER.	The *Fox and the Goose*,	Hevelius, 1690.	23	300°	25°

THE ZODIACAL CONSTELLATIONS.

No.	NAME.	MEANING.	BY WHOM ENUMERATED.	No. STARS.	R. A.	DEC.
1	ARIES,	The *Ram*,	Ptolemy.	17	37½°	18° N
2	TAURUS,	The *Bull*,	Ptolemy.	58	60°	18° "
3	GEMINI,	The *Twins*,	Ptolemy.	28	105°	25° "
4	CANCER,	The *Crab*,	Ptolemy.	15	130°	20° "
5	LEO,	The *Lion*,	Ptolemy.	47	155°	15° "
6	VIRGO,	The *Virgin*,	Ptolemy.	89	200°	3° "
7	LIBRA,	The *Balance*,	Ptolemy.	23	225°	15° S
8	SCORPIO,	The *Scorpion*,	Ptolemy.	84	244°	26° "
9	SAGITTARIUS,	The *Archer*,	Ptolemy.	88	285°	32° "
10	CAPRICORNUS,	The *Goat*,	Ptolemy.	22	315°	20° "
11	AQUARIUS,	The *Water-Carrier*,	Ptolemy.	25	330°	9° "
12	PISCES,	The *Fishes*,	Ptolemy.	18	5°	10° N

THE SOUTHERN CONSTELLATIONS.

No.	Name.	Meaning.	By whom enumerated or invented.	No. Stars.	R. A.	Dec.
1	Apis or Musca,	The *Bee or Fly,*	Bayer, 1604.	7	186°	68°
2	Ara,	The *Altar,*	Ptolemy.	15	256°	54°
3	Argo,	The *Ship Argo,*	Ptolemy.	133	115°	50°
4	Canis Major,	The *Great Dog,*	Ptolemy.	27	101°	24°
5	Canis Minor,	The *Lesser Dog,*	Ptolemy.	6	111°	5° N
6	Centaurus,	The *Centaur,*	Ptolemy.	54	195°	45°
7	Cetus,	The *Whale,*	Ptolemy.	82	30°	12°
8	Columba Noachi,	*Noah's Dove,*	Royer, 1679.	15	81°	35°
9	Corona Australis,	The *Southern Crown,*	Ptolemy.	7	277½°	40°
10	Corvus,	The *Crow,*	Ptolemy.	8	185°	18°
11	Crater,	The *Cup,*	Ptolemy.	9	170°	15°
12	Crux,	The *Cross,*	Royer, 1679.	10	184°	60°
13	Dorado,	The *Sword-Fish,*	Bayer, 1604.	17	70°	62°
14	Eridanus,	The *River Po,*	Ptolemy.	64	55°	30°
15	Grus,	The *Crane,*	Bayer, 1604.	11	335°	47°
16	Hydra,	The *Snake,*	Ptolemy.	49	150°	10°
17	Hydrus,	The *Water-Snake,*	Bayer, 1604.	25	40°	70°
18	Indus,	The *Indian,*	Bayer, 1604.	15	315°	55°
19	Lepus,	The *Hare,*	Ptolemy.	18	81°	20°
20	Lupus,	The *Wolf,*	Ptolemy.	34	231°	45°
21	Monoceros,	The *Unicorn,*	Hevelius, 1690.	12	105°	2°
22	Ophiuchus or Serpentarius,	The *Serpent Carrier,*	Ptolemy.	40	255°	0°
23	Orion,	The *Huntsman,*	Ptolemy.	37	82½°	0°
24	Pavo,	The *Peacock,*	Bayer, 1690.	27	290°	68°
25	Phœnix,	The *Phœnix,*	Bayer, 1690.	32	15°	50°
26	Piscis Australis,	The *Southern Fish,*	Ptolemy,	16	325°	32°
27	Touoan,	The *American Goose,*	Bayer, 1690.	21	356°	66°
28	Triangulum Australe,	The *Southern Triangle,*	Bayer, 1690.	11	235°	65°

b. **History of the Constellations.**—The following is a brief account of the origin of the constellations:

THE NORTHERN CONSTELLATIONS.

Andromeda, daughter of Cepheus, king of Ethiopia, who, to save his kingdom, caused her to be bound to a rock so that she might be devoured by a sea monster; but she was rescued by Perseus, who turned the monster into stone by presenting to it the head of Medusa, the Gorgon Queen, whom he had conquered and slain.

Aquila, according to the ancient fable, was king of one of the Cyclades, but was changed into an eagle and placed among the stars. Tycho Brahe added to this constellation *Antinous,* a youth of Asia Minor, greatly celebrated for his beauty, who lived in the reign of the Emperor Adrian.

QUESTION.—*b.* What is the history of each?

Auriga, represented as a man kneeling, and holding a bridle in his right hand, and a goat with her kids in his left. The accounts given of this constellation are various and inconsistent. Its origin is unknown.

Bootes, represented as grasping a club in one hand, while he holds the two hunting dogs by a cord, in the other. He seems to be driving the two bears round the pole. The origin of this constellation is lost in antiquity.

Canes Venatici, called *Asterion* and *Chara,* inserted by Hevelius, in 1690. They are held in a leash by Bootes.

Cassiopeia, the wife of Cepheus and mother of Andromeda.

Cepheus, king of Ethiopia, supposed to have gone with the Argonauts in search of the golden fleece.

Clypeus or **Scutum Sobieskii,** named in honor of John Sobieski, King of Poland, by Hevelius, who flourished during his reign.

Coma Berenices.—Berenice was the Queen of Ptolemy Euergetes, one of the kings of Egypt; and while he was engaged in war, she made a vow to dedicate her beautiful hair to Venus if he returned in safety,—a vow which she fulfilled.

Corona Borealis, a beautiful crown said to have been given by the god Bacchus to Ariadne, a Cretan princess.

Cygnus, supposed by some to represent the famous musician Orpheus, who, according to the fable, was changed into a swan, and placed near his lyre in the heavens.

Delphinus, the dolphin who is said, in the fable, to have persuaded Amphitrite to become the wife of Neptune, though she had made a vow of perpetual celibacy.

Draco, supposed to be the dragon which guarded the golden apples in the garden of the Hesperides, near Mount Atlas, and was slain by Hercules.

Equuleus is the head of a horse, supposed to have been the brother of Pegasus, and given to Castor by Mercury.

Hercules, the famous Grecian hero, celebrated for his many wonderful exploits.

Lyra, the harp given to Orpheus by Apollo, upon which he played with such skill, that even the rivers, it is fabled, ceased to flow in order to listen to his strains.

Pegasus, the winged horse, which, according to the Greek fable, sprung from the blood of Medusa, after Perseus had cut off her head. Bellerophon, attempting to fly to heaven upon his back, was dashed to the earth ; and the horse continuing his flight was finally placed by Jupiter among the constellations.

Perseus, son of Jupiter and Danae, was provided with celestial arms, and made war upon the three Gorgons, who had the power to turn every one to stone on whom they looked. Medusa was the most celebrated for her beauty ; but her hair was turned to serpents by Minerva, the sanctity of whose temple she had violated. (See Andromeda.)

Sagitta, the arrow of Hercules with which he killed the vulture that preyed on Prometheus, who was tied to a rock on Mount Caucasus by command of Jupiter.

Serpens and **Serpentarius,** or **Ophiuchus.**—The latter is supposed to be Æsculapius, a famous Greek physician. He holds the serpent in his hand as an emblem of his art, the cure of a serpent's bite being a test of medical skill.

Taurus Poniatowskii, a small constellation, some of the stars of which are arranged like a V, fancied to resemble a bull's head, and named as above after Count Poniatowski, who saved the life of Charles XII. King of Sweden, at the battle of Pultowa, and afterward at the battle of Rugen.

Triangulum.—This constellation represents the triangular delta of the Nile.

Ursa Major, supposed to represent Calisto, daughter of a king of Arcadia, and changed into a bear in consequence of the jealousy of Juno. *Ursa Minor* is supposed to have been her son Arcas, changed with her. These constellations are sometimes called *Triōnes ;* also, the *Greater* and *Lesser Wains.*

THE ZODIACAL CONSTELLATIONS.

These are supposed to have been invented by the Egyptians or Chaldeans to symbolize the changes of the months and seasons. The following is their origin according to the Greeks :

Aries is the ram that bore the golden fleece, for which the Argonauts undertook their expedition.

Taurus is supposed to be the bull whose form Jupiter assumed in order to carry off Europa, a beautiful princess of Phœnicia. She was borne across the sea to Europe, and gave name to that country. This constellation contains five stars arranged in the form of a V, and called the **Hyades.** It also contains the **Pleiades,** or *seven* stars, as the number is declared by some of the ancients, although only *six* are now visible.

Gemini, the twin brothers *Castor* and *Pollux*, sons of Jupiter and Leda, a Spartan queen. They were celebrated for their valor and heroic deeds, and were afterward worshiped as deities.

Cancer, the sea-crab, sent by Juno to annoy Hercules. This constellation more probably symbolizes the backward movement of the sun when at the northern solstice.

Leo, the famous Nemean lion, slain by Hercules; or a symbol of the intense heat of the season when this asterism is on the meridian.

Virgo, the Virgin Astræa, goddess of justice, who lived on earth during the golden age, but returned to heaven and was placed among the stars. A symbol, probably, of the time of harvest, as she holds an ear of corn (*spica virginis*) in her hand.

Libra, the *scales* which Virgo, the goddess of justice, used as an emblem of her office. Most probably it was an emblem of the balance or equality of the days and nights.

Scorpio, a very ancient constellation, supposed to typify the deadly influences of the season when the sun is in this part of the ecliptic. According to Ovid, it is the scorpion which stung Orion and caused his death.

Sagittarius, emblem of the hunters' season, inscribed on the Egyptian Zodiac. According to the Greeks, it represents *Chiron*, the famous Centaur, who changed himself partly into a horse.

Capricornus, emblem of the sun's climbing (as a goat) from the winter solstice. In the Greek fables, it is the goat into which Bacchus changed himself to escape the giant monster Typhon.

Aquarius, emblem of the wet season. The Greeks took it for Ganymede, the cup-bearer of Jupiter.

Pisces, emblem of the fishing season. The Greeks represent them to be the fishes into which Venus and Cupid changed themselves to escape the giant Typhon.

THE SOUTHERN CONSTELLATIONS.

Ara, the altar on which, according to the ancient mythology, the gods swore before their celebrated contest with the giants.

Argo, the ship in which the Argonauts sailed in quest of the golden fleece.

Canis Major and **Canis Minor**, supposed to be Orion's hounds.

Centaurus, one of the Centaurs, a fabulous race, half men and half horses ; most probably a tribe of men who invented, or were skilled in, the art of breaking in horses.

Cetus, the sea monster from which Andromeda was rescued by Perseus.

Corvus, the crow sent by Apollo to watch the conduct of his mistress Coronis, and rewarded by being placed in the heavens.

Crater.—The origin of this constellation appears to be unknown.

Crux, formed by Royer as a Christian emblem. It contains four stars that form a cross, two of them pointing directly to the south pole.

Eridanus, the river Po, fabled to have received Phaeton, who, having undertaken to guide the chariot of the sun, was struck by Jupiter with a thunderbolt, to prevent the general conflagration of the world from the ignorance of the rash youth.

Hydra, a monstrous serpent killed by Hercules. It is supposed by some to symbolize the moon's course.

Lepus, placed near Orion, as being one of the animals hunted by him.

Lupus, a king of Arcadia, *Lycaon*, changed into a wolf on account of his cruelties.

Ophiuchus.—See *Serpens.*

Orion, according to the Greeks, a famous hunter, who, as a punishment for his profane boasting, was bitten by the scorpion and killed. This constellation is mentioned in the book of Job, and is therefore of very great antiquity. Some think that it represents Nimrod, "the mighty hunter," mentioned in Genesis.

Piscis Australis, supposed by the Greeks to be Venus, transformed into a fish to escape the giant Typhon.

c. Of the 48 constellations enumerated by Ptolemy in his great work, the *Almagest,* all except three are described in a poem styled *Phenomena,* written in Greek by Aratus, a Cilician poet, 270 B.C., and still extant. This poem is, however, only a paraphrase of a celebrated work written about 370 B.C. by Eudoxus, a contemporary of Plato, and containing an account of all the constellations used in his time.

d. The following synopsis shows the relative positions as to right ascension and declination of the constellations visible in this latitude. If studied in this order they will be found without difficulty on the globe or in the heavens.

NOTE.—Each line in this table represents about 30° of right ascension.

TABLE SHOWING THE POSITIONS OF THE CONSTELLATIONS IN THE HEAVENS.

NORTH DECLINATION.				SOUTH DECLINATION.	
90°—50°	50°—25°	25°—0°	ZODIAC.	0°—25°	25°—50°
Cassiopeia	Andromeda Triangulum		*Pisces* *Aries*	Cetus	Phœnix
Camelopard.	Perseus		*Taurus*	{ Lepus Orion	{ Eridanus Columba
	Auriga	Canis Minor	*Gemini*	{ Canis Major Monoceros	Argo
	Lynx		*Cancer*		
Ursa Major	Leo Minor		*Leo*	{ Hydra Crater	
	Canes Ven.	Coma Ber.	*Virgo*	Corvus	Centaurus
Ursa Minor {	Bootes Corona Bor.	Serpens	*Libra*		Lupus
Draco	Hercules	Taurus Pon.	*Scorpio*	Ophiuchus	
	Lyra	{ Sagitta Aquila	*Sagittarius*	Clyp. Sobiesk.	Corona Aus.
Cepheus {	{ Lacerta Cygnus	{ Delphinus Equuleus Pegasus	*Capricornus* *Aquarius*		Grus

e. Each column of this table contains the constellations as they are arranged from west to east ; and each line read from *left* to *right,* gives the constellations, from north to south, that are on or near the meridian at the same time. By knowing the time that each zodiacal constellation comes to the meridian, remembering that these constellations are about 30° east of the signs, and that those are on the meridian at midnight which are opposite to the sun's place at noon, the student with a little consideration will be able to find the position of the constellations at any hour and on any evening during the year. The position at any time of the evening can easily be deduced from that at midnight by reckoning for each hour 15°, toward the east if the time is earlier, and toward the west, if later.

343. **Star Names.**—Only a very few of the stars are distinguished by particular names, the usual mode of designation being by means of letters or numbers.

The following is a list of the most noted or conspicuous of the stars with their special names, literal designations, magnitudes, and situations:

Note.—In the literal designations, the letter is followed by the Latin name of the constellation, in the possessive case; thus, α *Centauri* means the brightest star of Centaurus; β *Tauri*, the second star of Taurus; γ *Cassiopeiæ*, the third star of Cassiopeia, etc. The stars are arranged in this list according to their magnitudes.

LIST OF PRINCIPAL STARS.

Name.	Literal Designation.	Situation.	Magn.	R. A.	Dec.
Sirius	α *Canis Majoris*	Nose of the Great Dog	1	100°	16½° S
Canopus	α *Argûs*	The Ship Argo	1	95°	52½° S
Arcturus	α *Bootis*	Knee of Bootes	1	212°	20° N
Betelgeuse	α *Orionis*	Shoulder of Orion	1	87°	7½° N
Rigel	β *Orionis*	Foot of Orion	1	77°	8¼° S
Capella	α *Aurigæ*	Goat of Auriga	1	77°	46° N
Vega	α *Lyræ*	One of the strings of the Harp	1	278°	39° N
Procyon	α *Canis Minoris*	The Little Dog	1	113°	5½° N
Achernar	α *Eridani*	The River Po	1	23°	58° S
Aldebaran	α *Tauri*	Eye of the Bull	1	67°	16¼° N
Antares	α *Scorpionis*	Heart of the Scorpion	1	245°	26° S
Altair	α *Aquilæ*	Neck of the Eagle	1	300°	8½° N
Spica	α *Virginis*	Sheaf of Virgo	1	200°	10½° S
Fomalhaut	α *Piscis Aust.*	Southern Fish	1	343°	30¼° S
Regulus	α *Leonis*	Heart of the Lion	1	150°	12½° N
Deneb	α *Cygni*	Tail of the Swan	1	309°	45° N
Alpheratz	α *Andromedæ*	Head of Andromeda	1	½°	28½° N
Dubhe	α *Ursæ Majoris*	Great Bear	1½	164°	62½° N
Castor	α *Geminorum*	Heads of Gemini	1½	112°	32° N
Pollux	β *Geminorum*		2	114°	28¼° N
Pole-Star	α *Ursæ Minoris*	Tail of Little Bear	2	18½°	88¾° N
Alphard	α *Hydræ*	Heart of Hydra	2	140°	8° S
Ras Alhagus	α *Ophiuchi*	Head of Serpent-bearer	2	262°	12½° N
Markab	α *Pegasi*	Wing of Pegasus	2	345°	14½° N
Scheat	β *Pegasi*	Thigh of Pegasus	2	345°	27½° N
Algenib	γ *Pegasi*	Wing of Pegasus	2	2°	14½° N
Algol	β *Persei*	Head of Medusa	2½	45°	40½° N
Denebola	β *Leonis*	Tail of the Lion	2¼	176°	15° N
Alphecca	α *Coronæ Bor.*	Northern Crown	2½	232°	27° N
Benetnasch	η *Ursæ Majoris*	Tip of the Great Bear's Tail	2¼	206°	50° N
Alderamin	α *Cephei*	Breast of Cepheus	3	318°	62° N
Vindemiatrix	ε *Virginis*	Right Arm of Virgo	3	194°	11½° N
Cor Caroli	α *Canum Venaticorum*	The Hunting Dogs	3	193°	39° N
Alcyone	η *Tauri*	The Pleiades	3	55°	23¾° N

Questions.—343. To what extent are star-names used? Mention some of the principal.

PROBLEMS FOR THE CELESTIAL GLOBE.

PROBLEM I.—To find the place of a constellation or star on the globe: Bring the degree of right ascension belonging to the constellation or star to the meridian; and under the proper degree of declination will be the constellation or star, the place of which is required.

NOTE.—The student should be exercised in finding the places of all the constellations or stars laid down in the lists, according to this rule. The place of a planet or comet may also be found by this rule when its right ascension and declination are given.

PROBLEM II.—To find the appearance of the heavens at any place, the hour of the day and the day of the month being given : Make the elevation of the pole equal to the latitude of the place; find the sun's place in the ecliptic, bring it to the meridian, and set the index to 12. If the time be before noon, turn the globe eastward; if after noon, turn it westward till the index has passed over as many hours as the time wants of noon, or is past noon. The surface of the globe above the wooden horizon will then show the appearance of the heavens for the time.

NOTE.—The student must conceive himself situated at the centre of the globe looking out.

PROBLEM III.—To find the declination and right ascension of any constellation or star: Proceed in the same manner as to find the latitude and longitude of a place on the terrestrial globe.

STAR FIGURES

344. Particular stars can be easily recognized in the heavens by noticing the configurations which they form with each other, or by using the more conspicuous stars as "pointers;" that is, by assuming two bright stars so situated

that a straight line drawn through both will point directly
to the less prominent star whose position it is desired to
ascertain.

a. This is sometimes called the method of "alignments," and is
that usually employed by astronomers. A few examples are here
given in order to enable the student to find some of the most conspicuous of the stars.

1. **The Great Dipper, or Charles's Wain.**—This consists of seven
bright stars, in the Great Bear, so situated as to resemble a dipper
with a bent or curved handle, four of the stars forming the bowl, and
three, the handle.   It is situated within the circle of perpetual apparition, and hence is always visible, although in different positions as it
revolves around the pole.   The two stars at the far side of the bowl
(*a* and *β*) are called the "pointers" because a line drawn through them
would reach the pole-star, which can, therefore, always be found by
discovering the Great Dipper.   The pole-star, by modern astronomers
called *Polaris*, by the Greeks *Cynosura*, and by the Arabians *Alrucca-
bah*, is situated about $1\frac{1}{4}°$ from the pole, and forms the extremity of
the upwardly curved handle of a *small dipper*, which occupies a
reversed position from that of the Great Dipper, and consists of quite
faint stars.

2. **Trapezium of Draco.**—About 90° east of the Great Dipper are
four stars so arranged as to form an irregular quadrangle or trapezium.
These are in the head of Draco, and with another, a little to the west,
situated in the nose (*Rastaben*), form almost the letter V, pointing to
the west.

3. **The Chair of Cassiopeia.**—This consists of five stars of the 3d
magnitude, which, with one or two smaller ones, form the figure of an
inverted Chair; it is situated almost precisely at the opposite side of the
pole from the Dipper, being nearly 180° from it and in about the same
declination; it can thus be easily found.

4. **The Great Square of Pegasus.**— South of the Chair and a little to
the west, are four stars about 15° apart, forming a large square. They
are quite bright stars and the figure is very obvious. The north-eastern
star is *Alpheratz*, in the head of Andromeda; the south-eastern, *Algenib;* the south-western, *Markab;* and the north-western, *Scheat*.
Algenib and Alpheratz are on the equinoctial colure, which being continued toward the north passes through *Caph*, in Cassiopeia.

QUESTIONS.—*a.* What *star-figures* are described? What is the situation of each?

5. **The Great Y of Bootes.**—This consists of the bright and pecu-liarly ruddy star *Arcturus*, at the lower extremity of the letter; *Mirach*, at the fork; *Seginus*, at the extremity of the western arm; and *Alphecca*, in Corona Borealis, at that of the eastern. This figure is less than 45' to the south-east of the Great Dipper. *Arcturus* and *Seginus* form with *Cor Caroli*, situated toward the west, a large triangle; and a similar but a larger figure is also formed by *Arcturus* with *Dene-bola*, about 35° west, and *Spica Virginis*, about as far south.

6. **The Diamond of Virgo** is a large and very striking figure formed of *Cor Caroli* and *Spica Virginis*, at the extremities of its longest diagonal, and *Arcturus* and *Denebola* at those of the shortest. The former are about 50° apart; the latter 35½°. The figure extends from north to south.

7. **The Cross of Cygnus.**—This consists of five stars so arranged as to form a large and regular cross, the one at the upper extremity being *Deneb*, a star of the first magnitude. This figure is very mani-fest and is situated about 35° to the west of the Square of Pegasus. The star at the lower extremity of the cross is called *Albireo*. *Deneb*, or *Deneb Cygni* (*Deneb* means *tail*), is sometimes called *Arided*. A short distance toward the west from the *Cross* is the bright star *Vega*, forming with two faint stars near it a small triangle, the base being turned toward the side of the cross.

8. **The Sickle of Leo.**—If the line joining the " pointers " of the Great Dipper be continued toward the *south*, it will pass through a most beautiful object, having the complete form of a sickle, the bright star *Regulus* being at the extremity of the handle, and the curve of the blade toward the north-east.

9. **The V of Taurus.**—This is a group of stars situated in the head of the Bull, the brightest of which is Aldebaran, a ruddy star of the first magnitude, and situated at the left upper extremity of the letter. *Aldebaran* is an Arabic word and means, " that which follows:" it was applied to this star because it follows the Pleiades. This *group* of stars is called the *Hyades*. A little to the north-west is the famous cluster of small stars called the *Pleiades*, said once to consist of *seven* stars, although now we only discover *six*, of which *Alcyone* is the brightest.

10. **The " Bands of Orion."**—These are in a splendid group of stars to the south-east of Taurus, and situated under the equinoctial, consisting of four brilliant stars in the form of a long quadrangle

intersected in the middle by three stars arranged at equal distances in a straight line, and pointing to *Sirius*, the most splendid star in the heavens, on one side, and the *Hyades* and *Pleiades* on the other. These three stars have been called by some " The Yard ; " in the Book of Job they are called the *Bands of Orion*. A line of faint stars projects from these toward the south : these are sometimes called " The Ell." At the north-east extremity of the quadrangle is *Betelgeuse ;* at the south-east, *Saiph ;* at the south-west, *Rigel ;* and at the north-west, *Bellatrix.*

11. **The Crescent of Crater.**—To the south-east of the Sickle may be distinctly seen a beautiful crescent or semi-circle opening toward the west, consisting of stars of the sixth magnitude. They form the outlines of *Crater ;* and nearly south of the Sickle is the bright star *Cor Hydræ*, almost solitary in the heavens.

12. **The Dipper of Sagittarius.**—This is a very striking figure consisting of five stars of the 3d and 4th magnitudes, forming a straight-handled dipper turned bottom upward. It is a considerable distance south of Lyra, but comes to the meridian a very short time after it. The stars at the mouth of the dipper are about 5° apart.

Familiarized with these few configurations, it will not be difficult for the student, with the assistance of the globe or a planisphere, to acquire a knowledge of the other visible stars and their positions in the constellations to which they belong.

345. The APPARENT PLACES of the stars are constantly changing in consequence of the precession of the equinoxes. Their right ascensions and declinations are either increasing or diminishing, according to their situation, as the equinoctial pole revolves around that of the ecliptic.

a. The star *Polaris* is about 1½° from the pole, and is making a constant approach to it ; which it will continue to do until its distance is about ½°. It will then recede till in about 12,000 years the bright star *Vega*, which is now 51° 20′ from the pole will be less than 5° from it, and will therefore be the pole-star. About 4,000 years ago *a* Draconis was the polar star, being about 10′ from the pole.

346. NUTATION.—The precession of the equinoxes is not a uniform movement, but is subject to periodical variations

occasioned by the different positions of the sun and moon with respect to the plane of the equinoctial. When the sun is at the equinox its effect is nothing; at the solstice it is at its maximum; and thus arises, in connection with the general revolution of the pole, a vibratory motion of the earth's axis, called *nutation*.*

347. The SOLAR NUTATION is very slight and goes through all its changes in one year; but that of the moon, depending on the position of its nodes with respect to the earth's equinoxes, requires a period of 18½ years. The latter is what is ordinarily meant by nutation.

a. By the lunar nutation alone, the pole of the equator would be made to describe, in 18½ years. a small ellipse, about 18½" by 13¾", the longer axis being in the direction of the ecliptic pole; but being carried by the general movement of precession round the pole of the ecliptic at the rate of 50" annually, it actually moves in a circle the circumference of which is an undulating curve, somewhat like the real orbit of the moon, the limit of the undulation either way being 9¼".

b. The discovery of the nutation of the earth's axis was made by Dr. James Bradley in 1727, by noticing slight variations in the right ascensions and declinations of the stars of which neither precession nor any other known source of disturbance would account. The true cause of the phenomena soon suggested itself to his mind, but could not be confirmed until after 18½ years of observation. It was therefore not announced till 1745.

348. ABERRATION.—This is a change in the apparent places of the stars, which arises from the motion of the earth in its orbit, combined with the progressive motion of light.

a. This displacement of the stars was first observed by Hooke while attempting to discover a parallax in γ *Draconis*; but the true explanation of its cause was given by Dr. Bradley in 1727,—the year

---

* *Nutation* is derived from a Latin word which means a *nodding*.

---

in which the death of Newton took place. It was one of the most interesting and important astronomical discoveries ever made, and afforded an entire confirmation of the progressive motion of light, discovered by Roemer about 50 years previously.

b. **Cause of Aberration.**—An object is always seen in the direction in which the rays of light coming from it strike the eye. Now this depends not only on the actual direction of the rays themselves, but on our own motion with reference to them ; for if a ray is proceeding perpendicularly from an object and we are moving directly across it, it will appear to strike against the eye in an oblique direction, and thus the object will, in appearance, be thrown forward of its true place, by an angle depending for its size upon the ratio of the velocity of our own motion to that of light. This change of direction of the rays of light is similar to that which takes place in the drops of rain when we are running in a shower, and the rain descends perpendicularly ; for then it beats in our faces as it would if we were standing still and the wind were blowing it obliquely against us.

c. **Amount of Aberration.**—Since the velocity of light is 184,000 miles a second while that of the earth is but little more than 18 miles, the ratio of the latter to the former is about .0001, which is the sine of an angle of 20$\frac{1}{4}$'' ; and this accordingly is the amount of displacement due to aberration, when the star is so situated that the rays proceeding from it are perpendicular to the plane of the earth's orbit, the star in that case appearing each year to describe a small circle having a radius of 20$\frac{1}{4}$''. When the rays are oblique to this plane, the circle is foreshortened into an ellipse, and the amount of displacement varies, being 20$\frac{1}{4}$'' only when the rays are perpendicular to the earth's motion ; while in the case of stars situated in the plane of the ecliptic, there is merely an apparent oscillation, in a straight line, amounting to 41'' during each revolution of the earth.

d. The phenomena connected with aberration are thus very complicated ; and as they are all satisfactorily explained by the hypotheses of the earth's motion and that of light, both receive a confirmation from this important discovery.

349. THE GALAXY, OR MILKY WAY is that faint luminous zone which encompasses the heavens, and which, when

examined with a telescope, is found to consist of myriads cf stars. Its general course is inclined to the equinoctial at an angle of 63°, and intersects it at about 105° and 285° of right ascension. Its inclination to the plane of the ecliptic is consequently about 40°.

a. This nebulous zone is of very unequal breadth, not exceeding in some parts 3°; while in others it is 10° or even 16°; the average breadth being about 10°. It passes through Cassiopeia, Perseus, Gemini, Orion, Monoceros, Argo, the Southern Cross, Centaurus, Ophiuchus, Serpens, Aquila, Sagitta, Cygnus, and Cepheus. From *a* Centauri to Cygnus it is divided into two parts, the whole breadth including the two branches being about 22°. It exhibits other divisions at several points of its course.

350. Its appearance is not uniform, some parts being exceedingly brilliant; while others present the appearance of dark patches, or regions comparatively destitute of stars.

a. Near the Southern Cross, where its general appearance is most brilliant, there occurs a singular dark, pear-shaped space, obvious to the most careless observer. To this remarkable patch the early navigators gave the name of the *coal-sack.* A similar vacant space occurs in the northern hemisphere at Cygnus.

b. The number of stars in the **Milky Way** is inconceivably great. Sir William Herschel states that on one occasion he calculated that 116,000 stars passed through the field of his telescope in a quarter of an hour, and on another that as many as 258,000 stars were seen to pass in 41 minutes. The total number, therefore, can only be estimated in millions. Struve's estimate of the whole number visible in Sir William Herschel's great reflecting telescope is 20½ millions; and the number brought into view by the still more powerful instrument of Lord Rosse must be very much greater.

351. The PREVAILING THEORY with regard to the Milky Way is, that it is an immense cluster of stars having the general form of a mill-stone, split at one side into two folds, cr

thicknesses, inclined at a small angle to each other; that all the stars visible to us belong to this system; and that the sun is a member of it and is situated not far from the middle of its thickness, and near the point of its separation.

a. The fact that the Milky Way is composed of vast numbers of stars was conjectured by Pythagoras and other ancient astronomers, but was not positively discovered till Galileo directed his telescope to the heavens. The hypothesis that it is a vast cluster of which the sun and visible stars are members, was first suggested by Thomas Wright in a work entitled the "Theory of the Universe," published in 1750. This subject received a careful and prolonged investigation by Sir William Herschel, the results of which he published in 1784, and which seems to establish the hypothesis mentioned in the text. This opinion he arrived at by taking observations at different distances from the zone of the Galaxy, and counting the stars within the field of view. On the supposition that the stars are uniformly distributed throughout the system, the number thus presented would indicate the extent of the cluster in the direction in which they were seen ; and in this manner some general idea of its form would be obtained.

b. **Galactic Circle and Poles.**—The extensive survey made by Sir William Herschel of the stars in the northern hemisphere, and continued by his son, Sir John Herschel, in the southern, has proved that there are two points of the celestial sphere, diametrically opposite to each other, at which the stars are very thinly scattered ; while at and near the circle of which these are the poles the stars are so densely crowded as to be absolutely countless. This circle lies very near the middle course of the Milky Way, and hence is called the *Galactic Circle ;* the points at which the stars are least dense are called the *Galactic Poles.* It is also found that the decrease of the density of the visible stars in proceeding either way from the plane of the Galactic Circle conforms to the same law, but that the density in the southern hemisphere is at each latitude greater than at the corresponding latitude in the northern.

The annexed figure represents the general form of a section of this vast cluster, or stratum of stars, S being the place of the sun ; S *f,* the position of the plane of the Galactic Circle; *b b,* the Galactic Poles. It will be

---

obvious that at the visual lines Se and Sf the stars must appear most dense, and at Sb least; while at intermediate points, the density will vary with the obliquity of the visual lines; and as Sf is nearer the northern confines of the stratum than the southern, more stars must be visible in

Fig. 117.

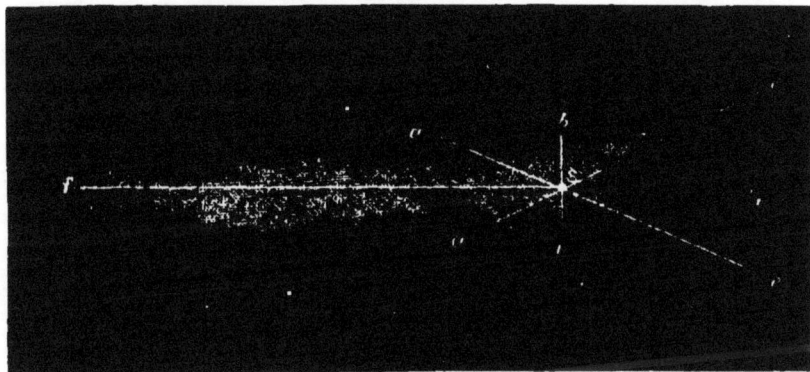

SECTION OF THE GALACTIC STRATUM.

the southern hemisphere than in the northern; the number of stars depending in each direction upon the length of the visual line. The apparent separation of the Milky Way is accounted for by supposing the sun to be placed, as indicated, near the point where the two branches diverge.

c. **Dimensions of the Galactic System.**—The thickness of this stratum of stars Herschel supposed to be about 80 times the distance of the nearest star from the solar system; but that its extreme length is equal to 2,000 times that distance. To move from one extreme point of this vast space to the other, light would require about 7,000 years.

352. PROPER MOTION OF THE STARS.—The stars do not always remain precisely in the same places with respect to each other, but in long periods of time perceptibly change their relative positions, some approaching each other, and others receding. This apparent change of position is called their *proper motion.*

a. The first astronomer to whom the idea of a proper motion of the stars (that is, a motion of the stars themselves, independent of annual parallax) occurred was Halley. Comparing the anciently recorded places of Sirius, Arcturus, and Aldebaran with their positions as observed by himself in 1717, and making every allowance for the variation in the obliquity of the ecliptic, he still found differences of latitude amounting to 37′, 42′, and 33′, respectively, for which he could not account, except on the supposition that the stars themselves had changed their positions. This was confirmed by Cassini in 1738, who ascertained that Arcturus had apparently moved 5′ in 152 years, while the neighboring star η Bootis had been nearly if not quite stationary. The star 61 Cygni has a considerable proper motion, having changed its position in fifty years nearly $4\frac{1}{2}′$.

b. **Motion of the Solar System in Space.**—In 1783 Herschel undertook the investigation of this interesting subject; and finding that in one part of the heavens the stars approached each other, while in the opposite part their relative distances seemed to increase, he arrived at the conclusion that this apparent change in the stars is caused by a real motion of the solar system in space. For, evidently, if we are in motion, the stars toward which we are moving will open out, while those from which we are receding will appear to come together; and as it was observed by Herschel that the stars in the constellation Hercules are gradually becoming wider and wider apart, he inferred that the motion of the sun and its attendant planets is in that direction. The mean result of the observations of Herschel, and several distinguished astronomers who in more recent times have investigated the subject, is that the point toward which the solar system is moving is in 260° 20′ of right ascension, and 33° 33′ of north declination, which agrees very nearly with that reached by Herschel himself. The annual angular displacement of a star situated at right angles to the direction of the sun's motion and at the mean distance of stars of the first magnitude, is computed at about $\frac{1}{4}″$; and therefore the velocity of the motion is estimated at about 160 millions of miles in a year.

c. **Central Sun.**—The hypothesis that the solar system is revolving around a central sun was first suggested by Wright in 1750. Mädler supposed that the central sun is the star *Alcyone* in the Pleiades; but it is not thought by astronomers that sufficient evidence exists for this hypothesis. All that can be said to be established is, that the sun with its great retinue of revolving worlds is moving in space toward a point in the constellation Hercules.

353. MULTIPLE STARS are those which to the naked eye

appear single, but when viewed through a telescope are separated into two or more stars. Those that consist of two stars are called *double stars*.

a. **Double stars** differ much in their distance from each other; in some cases being so near as to be separated only by the most powerful telescopes; in others, they are as much as ½′ from each other. These stars were carefully observed by Sir William Herschel, and have received much attention from the distinguished astronomers of more recent times. The list of this class of stars now contains upward of six thousand, classified according to their angular distances from each other.

b. The members of a double star are generally quite unequal in size. The pole-star consists of two stars of the second and ninth magnitude respectively, and about 18″ apart; Rigel has a companion star about 10″ from it, and of the ninth magnitude; Castor consists of two stars of the third and fourth magnitudes about 5″ apart; γ Virginis (*Gamma* of the Virgin) is a very remarkable star consisting of two stars each of the fourth magnitude. (See Fig. 118.) ε Lyræ (*Epsilon* of the Lyre) is an example

Fig. 118.

1. POLE-STAR; 2. RIGEL; 3. CASTOR; 4. γ VIRGINIS.

of a star consisting of two stars each of which is double, being thus a *double-double star*. In 1862, Sirius was discovered, by Mr. Alvan Clark, of New York, to have a minute companion star situated about 7″ from it.

c. **Colored Stars.**—There is considerable diversity in the color of both the single and double stars. Thus Vega, Altair, and Spica are white; Aldebaran, Arcturus, and Betelgeuse, ruddy; Capella and Procyon, yellow. Single stars of a fiery red or deep orange are not uncommon; but among the conspicuous stars there is only one instance (β Libræ) of a green star, and none of a blue one. Many

---

QUESTIONS.—*a.* Apparent distance of double stars? *b.* Comparative size and color? Size of the members of double stars? Examples? *c.* Difference in the color of single stars? Of double stars? Complementary colors? Presented by how many stars?

double stars exhibit the beautiful and curious phenomena of complementary colors,* the larger star being usually of a ruddy or orange
hue, and the smaller one, green or blue. In some cases, this is found
to be the effect of contrast; since, when a very bright object of a particular color is viewed with another less brilliant, the latter, although
in reality white, appears to have the complementary color of the former.
In this way a large and bright yellow object will cause other objects
to seem violet ; and crimson, produce the effect of green. In many
cases, however, there seems to be a real difference in the color of the
constituents of double stars ; for when one of them is concealed by
introducing a slide in the telescope, the other still retains its color.

Of 596 bright double stars contained in Struve's catalogue, 120 pairs
are of totally different hues. The number of reddish stars is double
that of the bluish stars ; and that of the white stars $2\frac{1}{2}$ times as great as
that of the red ones.

d. That some stars have changed in color is an established fact.
Ptolemy and Seneca expressly declare that Sirius was of a reddish hue ;
whereas now it is of a brilliant white. Stars described by Flamstead
were found by Herschel to have changed in this respect ; and γ Leonis
and γ Delphini have changed since his time.

354. BINARY STARS are double stars one of which revolves
around the other, or both revolve around their common centre of gravity.

a. **History of the Discovery.**—The discovery of this connection
between the constituents of double stars was, perhaps, the grandest of
Sir William Herschel's achievements. It was announced by him in
1803, after twenty-five years of patient observation, which he commenced with the view to discover the stellar parallax by noticing
whether any *annual* change in the relative positions of double stars
existed. To his astonishment, he found from year to year a regular
progressive movement of some of these bodies, indicating that they
actually revolve one round the other in regular orbits, and thus that

---

* *Complementary colors* are those which being blended produce white.
They are *red*, *yellow*, and *blue*. The complementary color of any one of
these is a combination of the other two. Thus *orange* is complementary of
blue ; and *green*, of red.

---

the law of gravitation extends to the stars. These stars are called *Binary* Stars*, or *Systems*, to distinguish them from other double stars which, although perhaps at immense distances from each other, appear in close proximity, because, as viewed from the earth, they are very nearly in the same visual line, and therefore are said to be *optically double.*

355. The observations of Herschel resulted in the discovery of about 50 binary stars; but since his time the number has been, according to Mädler, increased to 600. Most of the double stars are believed to be binary systems.

356. ORBITS AND PERIODS OF BINARY STARS.—A very careful scrutiny of these bodies and their changes in position has shown that they revolve in elliptical orbits of considerable eccentricity and in periods greatly varying in length.

The following is a list of the most remarkable of these bodies, with their periods, and the semi-axes and eccentricities of their orbits :

NAME.	PERIOD.	SEMI-AXIS MAJOR.	ECCENTRICITY.
	Years.	''	
ζ Herculis,	36.3	1.25	0.44
η Coronæ Borealis,	43.6	0.95	0.28
Sirius,	49.	7.05	
ζ Cancri,	58.9	1.29	0.23
ξ Ursæ Majoris,	63.1	2.45	0.39
α Centauri,	75 3	30.	0.96
ω Leonis,	84.5	0.85	0.64
70 Ophiuchi,	92.8	4.19	0.44
γ Coronæ Australis,	100.8	2.54	0.60
ξ Bootis,	117.1	12.56	0.59
δ Cygni,	178.7	1.81	0.60
η Cassiopeiæ,	181.	10.33	0.77
γ Virginis,	182.1	3.58	0.87
σ Coronæ Borealis,	195.1	2.71	0.30
Castor,	252.6	8.08	0.75
61 Cygni,	452.	15.4	
μ Bootis,	649.7	3.21	0.84
γ Leonis,	1200.		

* From the Latin word *bini*, meaning *two by two.*

It will be observed from the preceding table that the eccentricities of these orbits are as great as those of the comets.

b. **Dimensions of Stellar Orbits.**—In this table the semi-axis is given as seen perpendicularly from the earth; but to find the actual dimensions of the orbit, the parallax must be ascertained. The problem is then a simple one. Thus, the semi-axis of *a* Centauri is 30''; but since its parallax is 0.9187'', 1'' must at that distance subtend more than the semi-axis of the earth's orbit in the proportion of .9187 to 1 ; that is, it must be 1.088; and 30'' must subtend 1.088 × 30 = 32.64 times the semi-axis of the earth's orbit, which is equal to about 3,000 millions of miles. Now, the eccentricity is .96 ; and therefore the nearest distance to the central star is only .04, or 120 millions, while the farthest distance is 5,880 millions. In the case of 61 Cygni, the semi-axis is 15.4'', while the parallax is 0.3638'' ; hence, 1'' subtends at its distance 1 ÷.3638 = 2.75 (nearly); therefore the semi-axis 2.75 × 15.4 = 42.35 times the semi-axis of the earth's orbit, which is equal to 3,875 millions of miles.

c. *The following list contains all the stars whose parallax has been found :*

NAME.	PARALLAX	NAME.	PARALLAX
	''		''
a Centauri,	0.9187	*a* Lyræ,	0.155
61 Cygni,	0.5638	Sirius,	0.150
21258 Lalande,	0.2709	*ι* Ursæ Majoris,	0.133
17415 Oeltzen,	.247	Arcturus,	0.127
1830 Groombridge,	.226	Polaris,	0.067
70 Ophiuchi,	.16	Capella,	0.046

d. **Masses of the Stars.**—The joint mass, and in some cases the separate masses, of each pair of revolving stars can be ascertained, when we know their period and distance from each other. Thus, taking Sirius for example, we find its distance from its companion star to be 47 times the earth's distance from the sun, while its period is 49 times as great as the earth's. Hence, by the law stated in Art. 306, *a,* the

QUESTIONS.—*b.* Size of orbits—how found ? The calculation ? *c.* What is the nearest fixed star ? Parallax of Sirius ? What distance does it denote ? Capella ? *d.* Masses of the stars—how calculated ?

mass of the sun being 1, that of Sirius and its companion is $47^3 \div 49^2$ $= 43.25$. Now, it has been discovered* that Sirius is situated at a distance from the centre of gravity of both revolving stars equal to $16\frac{1}{4}$ times the earth's distance from the sun ; and therefore the companion star is $47 - 16\frac{1}{4} = 30\frac{3}{4}$ that distance; and as their masses are in inverse proportion to their distances from the centre of gravity, the mass of Sirius is to that of its satellite as $30\frac{3}{4}$ to $16\frac{1}{4}$, or as 123 to 65. Consequently, the mass of Sirius is $\frac{123}{188} \times 43\frac{1}{4} = 28.3$ times the mass of the sun ; and, if the densities are the same, its diameter is $\sqrt[3]{28.3}$, or a little more than three times that of the sun, and its disc 9 times as great. But photometric measurements have shown that its light is 400 times as great as that of the sun would be if the latter were removed to the distance of Sirius ; so that the materials of this star must be much less dense, or its light intrinsically far more brilliant, than that of the solar orb.

e. **The Sun a Small Star.**—By certain photometric comparisons recently made by Messrs. Clark and Bond between the star Vega (α Lyræ) and the sun, it has been shown that if the latter body were removed to 133,500 times its present distance, it would send us the same quantity of light as the star. But the nearest star (α Centauri) is more than 200,000 times as far from us as the sun ; and Vega, about six times as far as α Centauri. Hence the sun, if removed to the distance of the nearest star, would shine only as a star of the second magnitude ; and if removed to the mean distance of stars of the first magnitude, would appear as a star of the sixth magnitude, and be just visible to the naked eye. It would seem therefore that the sun, magnificent luminary as it appears to us, is only one of the smallest or least brilliant of the stars.

357. PHYSICAL CONSTITUTION OF THE STARS.—An analysis of the light of the stars indicates that they consist of solid incandescent matter surrounded with an atmosphere containing the vapor of some of the elementary substances existing on the earth ; such as mercury, antimony, sodium, hydrogen, etc.

a. **Spectrum Analysis.**—The band of rainbow colors, called the

---

* By Mr. Safford, of Chicago.

---

solar spectrum, produced by causing the sun's rays to pass through a piece of triangular glass called a prism, was noticed as early as 1802, by Wollaston, to be crossed by dark bands or lines; and in 1815, Fraunhofer, by examining the spectrum with a telescope, discovered as many as 500 of such lines, and since then the number perceived has increased to thousands. Now, it has also been observed that, when the light of any inflamed vapor passes through a prism, its spectrum consists of one or more bright-colored bands, differing in number, relative position, and color, according to the substance from which the vapor proceeds; but that when the light of any incandescent but *not vaporized* substance is made to pass through the inflamed vapor, the bright-colored lines are immediately changed to dark lines, the vapors absorbing from the light the same kind of rays which they themselves emit. Hence it is inferred that the substances whose peculiar lines are found in the solar spectrum are contained in a vaporous condition in the solar atmosphere, and as many as fourteen have been already identified. The stellar spectra also exhibit similar dark lines, each star having a peculiar series of them; and some are recognized as produced by the burning of substances found on the earth. Thus, some of the metals are found in some of the stars, and others in other stars; and this is thought to account for the different colors which the stars present. Sirius has been discovered to have five of our elements; and Aldebaran, nine.

358. Stars that appear double when viewed through an ordinary telescope are often separated by more powerful instruments, into *triple, quadruple,* or other *multiple* stars.

Fig. 119.

θ ORIONIS.   TRAPEZIUM OF ORION.

a. **Examples** are furnished by the following stars: ε Lyræ, already referred to, which consists of two stars, each of which is double; ζ Cancri (Zeta of the Crab), composed of three stars, two large and one small; θ Orionis (Theta of Orion), a very remarkable star, consisting of four bright stars, two of which have small companion stars, thus forming a *sextuple*

QUESTIONS.—358. What are triple stars, etc. *a.* Examples? Trapezium of Orion?

star. From the configuration of the four principal stars this is sometimes called the *trapezium of Orion.* (See Fig. 119.) As all these stars have the same proper motion, they are believed to constitute one system. It is said that a seventh star belonging to this system has been discovered by Mr. Lassell.

359. VARIABLE STARS are those which exhibit periodical changes of brightness. The number of such stars discovered up to the present time (1867) is about 120. They are sometimes called *Periodic Stars.*

a. **Examples.**—One of the most remarkable of these stars, and the first noticed (by Fabricius in 1596), is Mira—*the wonderful*—in the Whale (*o* Ceti). It appears about 12 times in 11 years; remains at its greatest brightness about a fortnight, being equal to a star of the 2d magnitude; decreases for about 3 months, and then becomes invisible, remaining so 5 months, after which it recovers its brillancy; the period of all its changes being about 331½ days.

Algol (*β* Persei) is another remarkable variable star of a very short period, it being only 2^d 20^h 49^m. It is commonly of the 2d magnitude, from which it descends to the 4th magnitude in about 3½ hours, and so remains about 20 minutes, after which in 3½ hours, it returns to the 2d magnitude and so continues 2^d 13^h, when similar changes recur. Observation shows that the period of Algol is less than it was in former years. Its variability was first noticed in 1669. *δ* Cephei is remarkable for the regularity of its period, which is 5^d 8^h 47^m. *Betelgeuse,* one of the four stars in the great quadrangle of Orion, has a period of 200 days. There is a star in Cygnus the variations of which are effected in 406 days. Three of the seven stars of the Great Dipper in Ursa Major are variable stars, their periods extending over several years. The double star *γ* Virginis is also variable, its two component stars having changed in brightness, the most brilliant becoming inferior to the other. *a* Cassiopeæ is also variable as well as double; and there are several others. According to Mr. Hind, the color of most variable stars is ruddy.

b. **Cause of Variable Stars.**—Several hypotheses have been suggested to account for these interesting phenomena. One is that these bodies rotate and thus present sides differing in brightness, or obscured

by spots similar to those which are seen on the solar disc ; another, that
their light is obscured by planets revolving around them ; and a third,
that their light is diminished by the interposition of nebulous masses,
since it has been observed that during their minimum brightness they
are often surrounded by a kind of cloud or mist. No one of these
hypotheses is entirely satisfactory, and hence we may conclude that
the true cause of the variability of these stars is unknown.

c. *The following is a list of the most interesting of these bodies :*

NAME.	PERIOD.	CHANGES OF MAGN.		NAME.	PERIOD	CHANGES OF MAGN.	
	Days.	From	to		Days.	From	to
β Persei,	2.86	2½	4	a Herculis,	88½	3	4
δ Cephei,	5.36	4	5	o Ceti,	331⅓	2	0
η Aquilæ,	7.17	3½	4½	v Hydræ,	449½	4	10
β Lyræ,	12.9	3½	4½	η Argûs,	46 yrs	1	4

360. TEMPORARY STARS are those which suddenly make
their appearance in the heavens, sometimes shining with
very great brilliancy; and, after a while, gradually fade
away, either entirely disappearing or remaining as faint
telescopic stars. The latter are properly called *New Stars.*

a. Ancient Instances.—The first on record was observed by Hipparchus, in the second century B.C. ; and it was the appearance of this
star that prompted him to make a catalogue of the stars,—the first
ever executed. This star seems to have been. noticed also by the
Chinese, as its appearance is mentioned in their chronicles under the
date of 134 B.C. Brilliant stars appeared in or near Cassiopeia in the
years 945 and 1264 A.D.

b. Star of 1572.—This was a very remarkable one, and is described
by Tycho Brahe, who observed it attentively. It appeared first as a
star of the first magnitude, blazing forth with the lustre of Jupiter or
Venus, and occasioning the greatest astonishment not only to scientific men, but to the common observers. For a while it was visible

even at noon. It lasted from November, 1572, to March, 1574,—17 months. Its color was successively white, yellow, red, and white again ; and its position in the heavens was the same during the whole time it remained visible. This star is supposed to be identical with those which appeared in 945 and 1264, all three being in fact apparitions of a variable star of a long period ; and some astronomers regard all temporary stars as of this character.

c. **Other Examples.**—In 1604 a very splendid star, remarkable for its vivid scintillation, shone forth in the constellation Ophiuchus, and lasted 15 months. This star was observed by Kepler and Galileo. In 1670 a star appeared in Cygnus, which attained the 3d magnitude and was visible for about two years, blazing out suddenly a short time before its final disappearance. On April 28th, 1848, a new star of the 5th magnitude was discovered in Ophiuchus, which in a few weeks rose to the 4th magnitude, but subsequently dwindled to the 12th magnitude, and still remains as a telescopic star. Lastly, a new star was seen in May, 1866, in Corona Borealis. It first appeared of the 2d magnitude, and of a pure white color ; but in a week had changed to the 4th magnitude, and a month afterward diminished to the 9th. " It is worthy of especial notice," says Sir John Herschel, " that all the stars of this kind on record, of which the places are distinctly indi-cated, have occurred in or close upon the borders of the Milky Way."

d. **Cause of Temporary Stars.**—No satisfactory hypothesis has as yet been advanced to account for these phenomena. Some have sup-posed that these stars are revolving in elliptical orbits of great eccentricity so that they sometimes approach very near us, and then recede to great distances ; but this is rendered improbable by the sud-den changes in brilliancy ; since, to pass from the first to the second magnitude, it is computed by Arago, would require six years, if the star moved with the velocity of light ; whereas, that of 1572 underwent this change in one month, and that of 1866 diminished to the extent of five magnitudes in the same time. Another hypothesis is, that extensive conflagrations take place on the surface of these bodies, which in their progress give rise to the observed changes in color and brightness, and 'at their extinction leave the body in an obscure state. The latter hypothesis has received some support from the recent inves-tigations made by Huggins, Miller, and others, by means of the spectrum analysis : for the light of the star of 1866 was shown by these experiments to proceed from matter in the state of luminous gas, chiefly hydrogen. Hence it is supposed that, by some great convulsion, large quantities of gas were evolved from the star, that the hydrogen was burning in combination with other elements, and that the inflamed gas had heated to incandescence the solid matter of the star. This hypothesis does not involve the necessity of destruction ; but, as remarked by Humboldt, only "a transition into new forms, determined by the action of new forces. Some stars which have become obscure, may again suddenly become luminous, by the renewal of the same conditions which, in the first instance, developed their light."

361. Numerous instances are on record of stars formerly known to exist which have entirely disappeared from the heavens. These are called *Lost or Missing Stars.*

a. **Examples.**—Several of the stars in the catalogue of Ptolemy were not to be found in 1433, when the catalogue of Ulugh Beigh was made at Samarcand ; and it is now known that 4 stars in Hercules have disappeared, 1 in Cancer, 1 in Perseus, 1 in Pisces, 1 in Hydra, 1 in Orion, and 2 in Coma Berenices. Sir William Herschel recorded in 1781 the star 55 Herculis ; but nine years afterward it was invisible, and has never been seen since. In 1670, it was remarked by Montanari, a distinguished astronomer, "There are now wanting in the heavens two stars of the 2d magnitude in the stern and yard of Argo. I and others observed them in 1664, but in 1668, not the least glimpse of them was to be seen."

b. **Why Stars Disappear.**—Some of the instances mentioned by early astronomers, of lost stars may be the result of erroneous entries ; but those of later times can not possibly be accounted for in this way. Revolving in orbits, they may have passed beyond the reach of the most powerful telescope ; or they be obscured by the interposition of great nebulous masses, and thus are only concealed for a certain period, which however may comprise hundreds, or even thousands of years.

362. STAR-CLUSTERS.—These are dense masses of stars so crowded together, and so far distant, that they present a hazy, cloud-like appearance, similar to that of the Milky Way.

QUESTIONS.—361. Lost stars? *a.* Examples? *b.* Why do stars disappear? 362. What are star-clusters?

a. Collections of stars visible as such to the naked eye, although considerably crowded, are called *star-groups.* Such are the *Pleiades,* the *Hyades,* and the group which constitutes the constellation Berenice's Hair. The first of these is a well-known object consisting of six stars when viewed by the naked eye, but exhibiting about 80 to the telescope. Seven of these stars have received special names, *Alcyone* being the brightest.

363. Among star-clusters, a very small number are sufficiently bright to be distinguished by the naked eye; but generally they require a telescope to render them visible.

a. Between the bright stars in Cassiopeia and P e r s e u s, there is a visible cluster, one of the most glorious objects in the heavens. (Fig. 120.) The remarkable group called *Præsepe,* or the " Beehive," is another example of a cluster visible without a telescope, but only as a spot of cloud It is situated in Cancer.

Fig. 120.

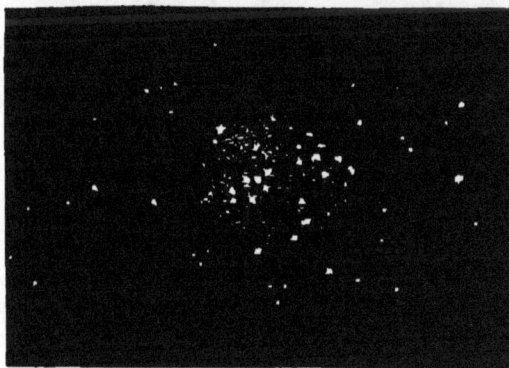

CLUSTER IN PERSEUS.

364. The PREVAILING FORM of the telescopic clusters is circular, with a gradual condensation of the luminous points toward the centre, indicating probably that the real form is that of a globe. Some of the clusters when viewed through powerful telescopes assume a much more irregular appearance, although their general form still appears to be spherical. Very irregular clusters are rare.

a. **Examples.**—A remarkable cluster surrounding the star Kappa

Fig. 121.

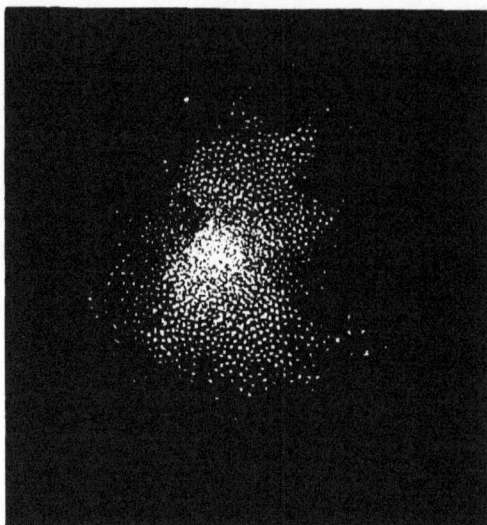

CLUSTER IN HERCULES.—*Sir J. Herschel.*

of the Southern Cross consists in part of colored stars; and another in the southern hemisphere is composed entirely of blue stars. The one situated between *Eta* and *Zeta* in the constellation Hercules is a peculiarly magnificent object in our northern heavens on fine nights, and is visible to the naked eye as a very small nebulous spot or faint star. In the telescope its general appearance is considerably changed by the addition of several outlying branches. This splendid object is represented in Fig. 121, in which its remarkable condensation at the central portions is strikingly exhibited. Very many objects of a similar character are visible in different parts of the heavens.

b. These globular clusters are supposed to be held together by their motions and mutual attractions. That there must be a real condensation is obvious from a simple glance at such an object as that depicted in Fig. 121; since the increase of brightness toward the centre is far too great to be explained on the supposition that the stars are equally distributed, but appear closer together at the centre, because the visual line traverses there a much greater portion of the mass.

c. The number of stars contained in these clusters is very great. According to Arago, many clusters contain at least 20,000 collected in a space, the apparent dimensions of which are scarcely a tenth as large as the disc of the moon. The clusters are not equally distributed over the heavens, but are most numerous in the Milky Way; while globular clusters most abound in that region of the Galaxy contained between Lupus and Sagittarius in the southern hemisphere.

QUESTIONS.—*b.* Cause of the globular form? Are the stars equally distributed? *c.* Number of stars in a cluster? Are the clusters equally distributed?

CHAPTER XIX.

365. NEBULÆ are certain faintly luminous appearances in the heavens, resembling specks of cloud or mist, some just visible to the naked eye, but the greater part only to be discerned with a telescope. They resemble in their general aspect the distant star-clusters, but their physical structure appears to be very different.

366. Their DISTANCE from us must be immense, since they constantly maintain very nearly the same situation with respect to each other and to the stars. Their magnitudes also must be inconceivably vast.

a. **History of their Discovery.**—It is only in quite recent years that any distinction could be positively made between nebulæ * proper and those immense star-clusters which present a nebulous appearance on account of their great distance. The first of these objects mentioned in the annals of astronomy was discovered in 1612, by Simon Marius, a German astronomer. This was the nebula situated in the girdle of Andromeda. In 1656, Huyghens discovered the great nebula in Orion, which he compared to an "opening in the heavens through which a brighter region beyond was visible." In 1716, Halley could enumerate only *six*, to which he added a few by his own discoveries; and during the next half century, the number was augmented to about 100. The labors of Sir William Herschel, directed to the investigation of this department of astronomy for more than twenty years, enabled him in

---

* *Nebula* is a Latin word meaning a *little cloud.* Plural, *nebulæ.*

1802 to publish a catalogue of 2,500 nebulæ and clusters; and the subsequent researches of his son, Sir John Herschel, in the southern hemisphere has increased this number to more than 5,000. Very great additions, however, have been made to our knowledge of these interesting objects by the labors of Lord Rosse, aided by the largest reflecting telescope ever constructed, and by the application of the spectrum analysis.

367. Nebulæ are distinguished from clusters by not being resolved into stars when viewed through the most powerful telescopes, presenting the appearance of diffuse luminous substances, filling vast regions of space, and differing in form, and degree of condensation.

a. Resolvable and Irresolvable Nebulæ. — Herschel at first thought all nebulæ resolvable into stars; but his subsequent investigations convinced him that this was an error; and he accordingly divided these objects into *resolvable and irresolvable nebulæ ;* the first being those vast star-clusters which exhibit a nebulous, or cloudy aspect, because of their comparatively crowded condition and great distance from us; and the second, according to his conceptions, *immense aggregations of self-luminous matter,* of great tenuity, but gradually condensing into solid bodies like the sun and stars. This bold conception of Herschel's had been entertained by Tycho Brahe and Kepler. who suggested that the new stars seen in their time were caused by aggregations of the ethereal matter filling space. The first discoverers of the nebulæ also noticed an essential difference between the light of these great phosphorescent masses and that of the stellar clusters, fancifully comparing the former to glimpses of the empyrean disclosed by rents or chasms in the celestial vault. Laplace applied this conjecture of Herschel's, under the name of the " Nebular Hypothesis " to the solar system, in order to explain the manner of its evolution. (See page 202.)　Many of the irresolvable nebulæ of Sir William Herschel having been resolved by the great telescope of Lord Rosse, or having given indications of being resolvable into stars, the opinion came to be almost universally entertained that *all nebulæ* are star-clusters, some so distant that light requires millions of years to pass from them to us. But the spectrum analysis has proved this to

---

be erroneous, by showing that these luminous masses consist of *gaseous*, not *solid* matter; so that Herschel's hypothesis would seem to be established. These diffuse and attenuated substances constitute thus a peculiar class of objects in the starry heavens, and are the *nebulæ* defined in the text, although some astronomers still continue to classify them with the clusters which have a nebulous appearance.

368. Nebulæ abound chiefly in those regions of the heavens in which the stars are least numerous, that is, in the vicinity of the Galactic Poles, and are more uniformly spread over the zone which surrounds the South Pole. Where stars are excessively abundant, nebulæ are rare.

a. Herschel found this rule to be without exception; and whenever, during a brief interval, no star passed into the field of his telescope, as in the diurnal motion the heavens swept by it, he was accustomed to say to his secretary, "Prepare to write : nebulæ are about to arrive."

369. Nebulæ may be divided, according to their form, into the following six classes; namely, *elliptic, annular, spiral, planetary, stellar,* and *irregular nebulæ.*

370. ELLIPTIC NEBULÆ are such as have the elliptical or oval form. They are quite numerous and are of various degrees of eccentricity.

Fig. 122.

ELLIPTIC NEBULA IN AN-
DROMEDA.

a. **Examples.** — A remarkable nebula of this kind is represented in Figure 122. It is situated in the right foot of Andromeda. The most noted of these nebulæ is that in the girdle of Andromeda, already referred to as the first nebula discovered. It is one of the grandest and least resolvable in the heavens, and is visible to the naked eye; so much so, indeed, that it is strange none of the ancient astronomers ever observed it. In very powerful telescopes, such as that at Cambridge, Mass., it presents some quite striking peculiarities of

QUESTIONS.—368. Where are nebulæ abundant? *a.* Herschel's remark? 369. How are nebulæ classified? 370. Elliptic nebulæ? *a.* The nebula in Andromeda.

form, two dark clefts appearing to divide it longitudinally ; while its length, as viewed through this instrument by Prof. Bond, is 4°, and its breadth, 2½°.

Fig. 123.

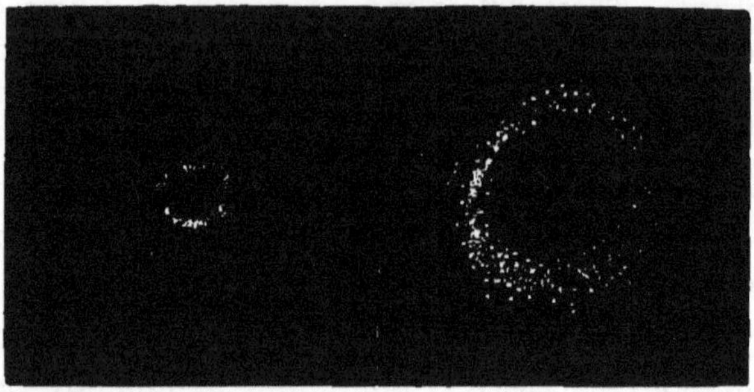

ANNULAR NEBULA IN LYRA.—1. *Sir John Herschel ;* 2. *Lord Rosse.*

371. ANNULAR NEBULÆ are such as have the form of a ring. These are very rare, the heavens affording only four examples.

a. The most remarkable one is found in Lyra, situated between the stars *Beta* and *Gamma*, and may be seen with a telescope of moderate power. It is slightly elliptical and has the appearance of a flat oval ring, the opening occupying somewhat more than one-half of the diameter. The central portion is not altogether dark, but is crossed with faint nebulous streaks, compared by some to gauze stretched over a hoop. The telescope of Lord Rosse shows fringes of stars at its inner and outer edges. (See Fig. 123.) The other annular nebulæ are two in Scorpio, and one in Cygnus.

372. SPIRAL NEBULÆ are such as have the form of one or more spirals or coils ; in some cases presenting the appearance of continuous convolutions, or whorls ; in others, of great spiral arms or branches projecting from a central nucleus.

a. The discovery of nebulæ of this remarkable form is due to Lord Rosse, no indication of it whatever having been afforded by the great

QUESTIONS.—371. What are annular nebulæ ? Their number ? *a.* The most remarkable ? 372. What are spiral nebulæ ? *a.* By whom discovered ? The most remarkable ? Describe the one in Leo.

telescope of Sir William Herschel. The grandest object of this kind
is found in Canes Venatici. Brilliant spirals, unequal in size and
brightness, and apparently overspread with a multitude of stars,
diverge from the central nucleus, the whole suggesting the idea of a
rotary movement of considerable rapidity, and the play of forces at
which the imagination is startled when it contemplates the immensity
of space filled by this wondrous object.

Fig. 124.

SPIRAL NEBULÆ IN LEO.—*Lord Rosse.*

Figure 124 represents a very beautiful object of this kind in the lower
jaw of Leo, the spiral form being clearly brought out in Lord Rosse's great
telescope. The convolutions are nearly all closed, so as to assume almost
the form of concentric ellipses, the central one containing what appear like
several distinct stellar nuclei.

373. PLANETARY NEBULÆ are those which, in their circular or slightly elliptical form, their pale and uniform light,
and their definite outline, resemble the larger and more distant planets of our system.

a. One of the most striking of this class is found in Ursa Major
(near β), the light of which, in Sir John Herschel's drawing, is quite
uniform ; but when seen through Lord Rosse's telescope, it presents
the appearance depicted in Fig. 125 (No. 1). The disc is about 3' in
diameter, and exhibits a double luminous circle with two dark openings,

each containing a bright but partially nebulous star. No. 2 in the same figure, represents a nebula near κ (*Kappa*) in Andromeda, which, though perfectly round in Herschel's drawing, appears in Lord Rosse's like a luminous ring surrounded by a wide nebulous border.

Fig. 125.

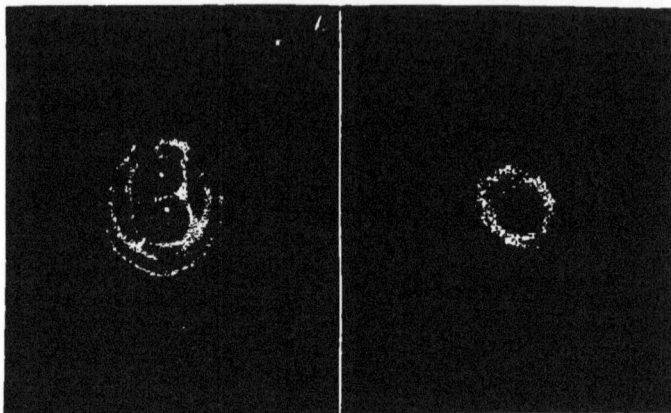

PLANETARY NEBULÆ.   1, IN URSA MAJOR , 2, IN ANDROMEDA.

b. The number of planetary nebulæ discovered is about 25, three fourths of which are in the southern hemisphere. Several described by Sir John Herschel are of a blue color, in some cases with a tinge of green.

c. The size of these objects must be amazingly great. That of Ursa Major, if no farther from us than the nearest star, α Centauri, would be sufficiently large to fill a space equal to three times the orbit of Neptune ; but there is reason to believe that it is more than 1,000 times as large as this. How vast then must be the size of such a nebula as that in Andromeda !

In Fig. 126, 1 and 2 are representations of planetary nebulæ. The former is in Aquarius and is quite remarkable for its brightness.

374. STELLAR NEBULÆ are those which appear to envelope one or more brilliant spots or points, resembling stars surrounded by a nebulous border or ring. There are several varieties, the most important of which are *nebulous stars.*

375. NEBULOUS STARS are stars encircled by a nebulous border, which in some cases has a clearly defined outline, in others gradually shading off into the general appearance of the sky.

Fig. 126.

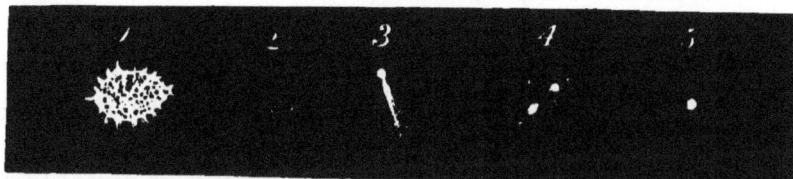

a. Such stars differ from other stars only in having this apparently nebulous atmosphere. If the nebula is circular the star occupies the centre of it; while in the case of some that are elliptical, two stars are placed at the foci.

In the above cut (Fig. 126), No. 5 represents a remarkable nebulous star in Cygnus. The star is of the 11th magnitude, and is at the centre of a perfectly circular nebula of uniform light, and about 15' in diameter. No. 4 is a stellar nebula in Sobieski's Shield, of an elliptic form, and having two stars at the foci of the ellipse. These stars are described by Sir John Herschel as of a gray color. No. 3 is the representation of a nebula bearing a resemblance to a comet. It is found in the tail of Scorpio. There are several other instances of such nebulæ, which from their appearance are called *conical* or *cometary nebulæ.* In the case of each the stellar, or bright, point is at one extremity of the nebulous mass.

b. It has been thought by some that the connection between these nebulæ and stars is not real, but is merely the effect of perspective, the one being situated behind the other, but separated by a wide interval. It is, however, very improbable that so many of these nebulæ should be found with stars placed exactly at their centres, some gradually becoming fainter towards their borders, if there were no physical connection between them; especially as, up to the present time, no difference in their proper motion has been discovered.

c. It seems, therefore, probable that these are stars encompassed with very extensive atmospheres in the same way, perhaps, as the luminous centre of our system is enveloped in what we call the Zodical

Light, the sun itself being in reality a nebulous star; but the atmospheric envelopes of the other stars referred to must be vastly more extensive. The one in Cygnus, described above as 15′ in diameter, and therefore extending 450″ beyond the central star, must, if the object is only as far from us as α Centauri, have an extent equal to fifteen times the distance of Neptune from the sun.

376. Irregular Nebulæ are such as have no symmetry of form and scarcely any distinctness of outline, and are also remarkable for the diversity of brightness which they exhibit at different parts.

a. Arago remarks of these diffuse masses of nebulous matter, that "they present all the fantastic figures which characterize clouds carried away and tossed about by violent and often contrary winds." The most remarkable of these objects are the following:

Fig. 127.

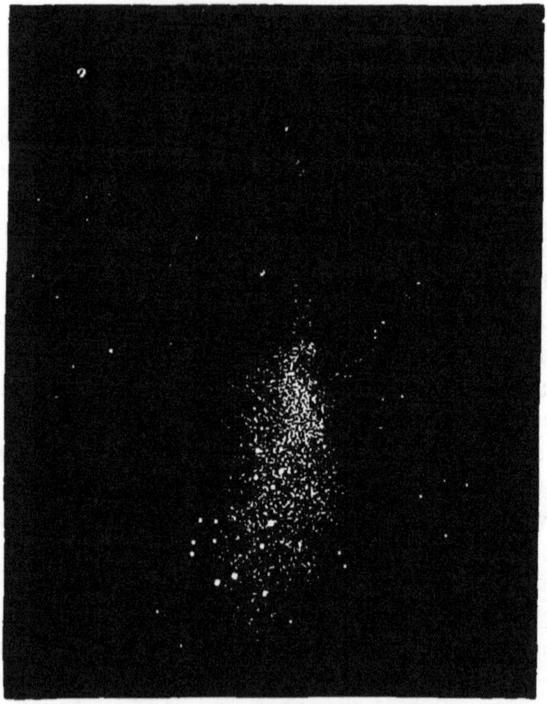

CRAB NEBULA IN TAURUS.—*Lord Rosse.*

Questions.—376. What are irregular nebulæ? *a.* Their general appearance?

1. *The Crab Nebula in Taurus* (Fig. 127).—This singular object has an elliptic outline in ordinary telescopes, but in Lord Rosse's great reflector it presents an appearance which has been fancifully likened to a crab or lobster with long claws.

2. *The Great Nebula in Orion.*—This is probably the most magnificent of all the nebulæ. It is very irregular in form ; of immense extent, covering a surface more than 40' square ; and consists of patches varying considerably in brightness. Near the famous sextuple star θ Orionis, already described, it is very brilliant ; but other portions are quite dim, and some absolutely black. It was thought that portions of this nebula had been resolved into stars by the telescopes of Lord Rosse and Prof. Bond ; but the experiments of Messrs. Huggins and Miller with the spectroscope have proved conclusively the gaseous nature of this object ; the light from the brightest part of the nebula giving a spectrum of only three bright lines, indicating the presence of *hydrogen*, *nitrogen*, and a third substance unknown. The examination of other portions gave similar results.

3. *The Great Nebula in Argo.*—This is another very irregular and extensive nebula, covering a space equal to five times the disc of the moon. It contains a singular vacancy of an irregular oval form near the centre, and not very far from the variable star *Eta* " It is not easy," says Sir J. Herschel, " to convey a full impression of the beauty and sublimity of the spectacle which this nebula offers as it enters the field of the telescope, ushered in as it is by so glorious a procession of stars, to which it forms a sort of climax." This nebula is remarkably destitute of any indications of resolvability.

4 *The Dumb-bell Nebula* (Fig. 128.)—This object is found in Vulpecula, and derives its name from its singular appearance as viewed through a telescope of moderate power. In Lord Rosse's telescope it assumes a form of less regularity, and appears to consist of in-

Fig. 128.

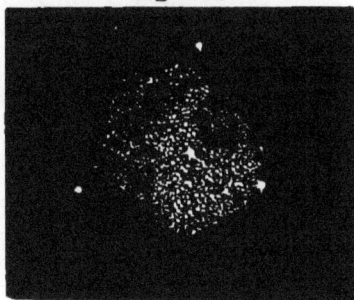

DUMB-BELL NEBULA.—*Herschel.*

numerable stars mixed with a mass of nebulous matter. These may be only centres of condensation.

5. *The Magellanic Clouds.*—These are situated in the southern hemisphere and not far from the pole, and are called sometimes *Nubecula Major and Minor*, or the *Greater and Lesser Cloudlets*. The former is in Dorado; the latter in Toucan. These objects are distinguished for their great extent, the larger one covering a space of about 42 square degrees, and the smaller being of about one-fourth that extent, but of greater brightness. The telescope of Sir J. Herschel decomposed them into separate stars, star-clusters, and numerous distinct nebulæ. In the larger cloud, Herschel counted 582 single stars, 46 star-clusters, and 291 nebulæ, and in the smaller cloud, 200 single stars, 7 star-clusters, and 37 nebulæ. In the immensity of their extent and the diversity of objects which they present, they are only comparable to that apparently greatest of all clusters, the Milky Way.

377. DOUBLE NEBULÆ are those which indicate by their close proximity to each other that they have a physical connection. More than 50 of such objects have been enumerated, the component nebulæ of which are not more than 5' apart.

Fig. 129 represents an object of this kind, found in Gemini. It is composed of two rounded masses, terminated by brilliant appendages and enveloped in a nebulous mass, the whole surrounded by light luminous arcs resembling fragments of a nebulous ring.

Fig. 129.

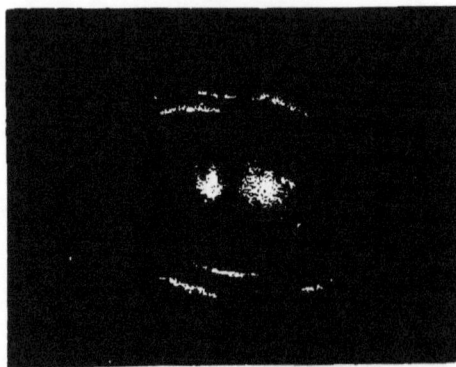

DOUBLE NEBULA.—*Lord Rosse.*

378. VARIABLE NEBULÆ are those which undergo changes in apparent form and brightness.

a. Several instances of such changes have been positively ascertained by Struve, D'Arrest, Hind, and other distinguished astronomers. The

great nebulæ in Orion and Argo have exhibited undoubted variations of a marked character. When Sir J. Herschel observed the latter
in 1838, the star *Eta* was of the 1st magnitude and situated in the
densest part of the nebula, but in 1867, an observer at Hobart-town,
found it within the central vacuity, and only of the 6th magnitude.
It is also stated that the vacuity is materially different in form from
that represented by Herschel. Another observer at Madras confirms
these statements, and adds that the nebula has varied considerably in
brightness while under his own observation.

Some of the smaller nebulæ exhibit changes similar to those of the
variable stars. In 1861, it was noticed by D'Arrest that one in Taurus had disappeared; and circumstances indicated that this event had
occurred in 1858. In December, 1861, the nebula reappeared, increased
in brightness for several months, but in December, 1863, could not be
found. Phenomena of a similar character were observed by Sir W.
Herschel. Two stars, surrounded by circular nebulæ in 1774, presented
no traces of these envelopes in 1811. If these objects, as it was formerly
supposed, were all composed of distinct stars, it would be scarcely possible to conceive how such variations or disappearances could occur,
particularly within the short periods mentioned; but on the hypothesis that they consist of diffuse luminous matter, each nebula being a
separate mass, these changes harmonize with what we see among the
stars themselves.

379. STRUCTURE OF THE UNIVERSE.—The universe has
been supposed, by many modern astronomers, to consist of
an infinite number of star-clusters similar to the galaxy, and
situated at inconceivably immense distances from it and
from each other. In view, however, of the recent discoveries as to the nature of the nebulæ proper, this hypothesis
can not be considered as established; and the true structure
of the universe remains a problem to be solved.

a. The hypothesis alluded to was a deduction from that which, supposing every nebula to be resolvable into stars, banished those that
seemed irresolvable to the uttermost depths of space. Spectrum analysis having exploded this idea, we are necessarily compelled to
discard those extravagant conceptions as to the distance of these visible

objects. For, since we can not penetrate to the remotest parts of the galaxy, or resolve every portion of its milky light into stars, there is no reason for believing that those star-clusters, which are readily resolvable, are beyond the confines of our sidereal system; while the fact, already mentioned, that clusters and nebulæ are invariably abundant where stars are rare, and as invariably wanting where stars abound, affords presumptive evidence that all these bodies are physically con-nected with the same great system of the universe of which the galaxy itself is a portion.

b. What other creations occupy the infinitude of space beyond the reach of human vision aided by the utmost efforts of optical and me-chanical skill, we can neither know nor perhaps conceive. There is reason for believing that light itself is gradually absorbed and thus extinguished in its journeyings from those remote regions of the uni-verse, long before it could reach our little orb and give us intelligence of the worlds from which it sped. But that the works of God are infinite in extent as they are in perfection and beneficent design, we can not but believe; nor as we contemplate the wonders and glories of the starry heavens—those unfathomable abysses lit up by millions of suns, can we refrain from bowing in adoration and gratitude to Him who has endowed us with the intellectual power (far more won-drous than even these worlds themselves) to discover and survey their vastness and magnificence, and with those moral and spiritual capacities, by the due cultivation of which we may prepare ourselves for an existence in that future world where we shall be enabled, in a far higher degree, to contemplate His power and to understand His infinite wisdom and beneficence.

QUESTIONS.—*b.* Other creations in the infinitude of space ? Why not discoverable ? Feelings excited by a contemplation of the starry heavens ?

APPENDIX.

TABLE I.—ELEMENTS OF THE SOLAR SYSTEM—(SUN's PARALLAX, 8.94").

Name.	Sign.	Mean Diameter.	Oblateness.	Mass, ⊕ being 1.	Density compared with water.	Mean Distance in Millions.	Eccentricity of Orbit.	Inclination of Orbit.	Sidereal Period.	Synodic Period.	Time of Rotation.	Inclination of Axis.
									yrs. dys.	dys.		
SUN.........	☉	852,900		315,000.	1.42						25 d. 8 h.	7° 20'
MERCURY. .	☿	2,962	?	.065	6.35	35.4	.205	7°	88	116	24 h. 5 m.	?
VENUS.....	♀	7,510	?	.885	5.84	66.15	.0009	3¼°	224¾	584¼	23 h. 21 m.	75°
EARTH.....	⊕	7,912	¹⁄₂₉₉	1	5.67	91.5	.017		365¼		24 h.	23° 28'
MARS.......	♂	4,300	¹⁄₅₀?	.118	3.97	139.3	.093	1° 51'	1 322	780	24 h. 37 m.	28° 42'
JUPITER....	♃	85,000	¹⁄₁₇	301.	1.37	475.75	.048	1° 19'	11 315	399	9 h. 55½ m.	3° 6'
SATURN....	♄	70,100	¹⁄₁₀	90.	.74	872.	.056	2° 30'	29 167	378	10 h. 29 m.	26° 49'
URANUS....	♅	33.247	¹⁄₁₀?	12.65	.97	1,754.	.047	46½'	84 6	369½	?	?
NEPTUNE...	♆	36,806	?	16.8	.91	2,746.	.0087	1° 47'	164 226	367½	?	?
MOON......	●	2,162	?	¹⁄₈₀	3.4	.2388	.055	5⅛°	27½	29¼	27¾ d.	6° 39'

TABLE II.—ELEMENTS OF THE MINOR PLANETS.

NAME.	Number.	Mean Distance.	Eccentricity.	Inclination of Orbit.		Sidereal Period.		Discoverer.	Date.
		⊕'s=1		°	′	Yrs.	dys.		
FLORA	8	2.2014	.157	5	53	3	97	Hind	1847
ARIADNE	43	2.2034	.168	8	28	3	99	Pogson	1857
FERONIA	71	2.2661	.12	5	24	3	150	C. H. F. Peters.	1861
HARMONIA	40	2.2677	.046	4	16	3	151	Goldschmldt...	1856
MELPOMENE	18	2.2956	.217	10	9	3	174	Hind	1852
SAPPHO	80	2.2963	.2	8	37	3	175	Pogson	1864
VICTORIA	12	2.3344	.219	8	23	3	207	Hind	1850
EUTERPE	27	2.3467	.173	1	35	3	217	Hind	1853
VESTA	4	2.3733	.09	7	8	3	229	Olbers	1807
URANIA	30	2.3655	.126	2	6	3	223	Hind	1854
NEMAUSA	52	2.3657	.066	9	57	3	223	Laurent	1858
CLIO	84	2.3675	.238	9	22	3	225	Luther	1865
IRIS	7	2.3862	.231	5	28	3	240	Hind	1847
METIS	9	2.3866	.123	5	36	3	251	Graham	1849
ECHO	62	2.393	.185	3	34	3	256	Ferguson	1860
AUSONIA	63	2.395	.126	5	47	3	258	De Gasparis	1861
PHOCEA	25	2.4008	.254	21	35	3	263	Chacornac	1853
MASSILIA	20	2.4097	.144		41	3	270	De Gasparis	1852
ASIA	67	2.4217	.185	5	59	3	280	Pogson	1861
NYSA	44	2.422	.151	3	42	3	281	Goldschmidt	1857
HEBE	6	2.4259	.203	14	47	3	284	Hencke	1847
BEATRIX	83	2.4287	.084	5	2	3	287	De Gasparis	1865
LUTETIA	21	2.4354	.162	3	5	3	292	Goldschmidt	1852
ISIS	42	2.44	.225	8	34	3	296	Pogson	1856
FORTUNA	19	2.4411	.158	1	33	3	297	Hind	1852
EURYNOME	79	2.4431	.195	4	37	3	299	Watson	1863
PARTHENOPE	11	2.4519	.099	4	37	3	306	De Gasparis	1850
THETIS	17	2.4735	.128	5	36	3	325	Luther	1852
HESTIA	46	2.5265	.164	2	18	4	6	Pogson	1857
	89	2.5498	.18	16	11	4	26	Stephan	1866
AMPHITRITE	29	2.554	.074	6	8	4	30	Marth	1854
EGERIA	13	2.5766	.087	16	31	4	50	De Gasparis	1850
ASTRÆA	5	2.5771	.187	5	19	4	51	Hencke	1845
IRENE	14	2.586	.166	9	8	4	58	Hind	1851
POMONA	32	2.5873	.083	5	29	4	59	Goldschmidt	1854
	91	2.5958	..					Stephan	1866
MELETE	47	2.5959	.287	8	1	4	67	Goldschmidt	1857
PANOPEA	70	2.6133	.183	11	88	4	82	"	1861
CALYPSO	54	2.6197	.204	5	7	4	88	Luther	1858
DIANA	78	2.6228	.205	8	88	4	90	"	1863
THALIA	23	2.6271	.232	10	13	4	94	Hind	1852
FIDES	37	2.6414	.177	3	7	4	107	Luther	1855
EUNOMIA	15	2.6437	.187	11	44	4	109	De Gasparis	1851
VIRGINIA	51	2.6491	.287	2	48	4	114	Ferguson	1857
MAIA	66	2.6512	.158	8	4	4	116	H. P. Tuttle	1861
IO	85	2.6536	.191	11	53	4	118	Peters	1865
PROSERPINE	26	2.6561	.087	3	36	4	120	Luther	1853
CLYTIE	73	2.6666	.043	2	25	4	129	Tuttle	1862
JUNO	3	2.6684	.257	13	1	4	131	Harding	1804
EURYDICE	75	2.6698	.307	5	0	4	133	Peters	1862
FRIGGA	77	2.6719	.136	2	28	4	134	"	"

ELEMENTS OF THE MINOR PLANETS — CONTINUED.

NAME.	Number.	Mean Distance.	Eccentricity.	Inclination of Orbit.		Sidereal Period.		Discoverer.	Date.
		⊕'s=1.		°	′	Yrs.	dys.		
ANGELINA	64	2.6809	.128	1	20	4	142	Tempel	1861
CIRCE	34	2.6863	.167	5	26	4	147	Chacornac	1855
CONCORDIA	58	2.7008	.042	5	2	4	160	Luther	1860
ALEXANDRA	55	2.7123	.197	11	47	4	171	Goldschmidt	1858
OLYMPIA	60	2.7131	.117	8	87	4	172	Chacornac	1860
EUGENIA	45	2.7212	.08	6	85	4	179	Goldschmidt	1857
LEDA	38	2.7401	.155	6	58	4	196	Chacornac	1856
ATALANTA	36	2.7461	.301	18	42	4	201	Goldschmidt	1855
NIOBE	72	2.7554	.174	23	19	4	209	Luther	1861
PANDORA	56	2.7591	.145	7	14	4	213	Searle	1858
ALCMENE	82	2.7608	.226	2	51	4	214	Luther	1864
CERES	1	2.7667	.08	10	86	4	220	Piazzi	1801
LÆTITIA	39	2.7671	.115	10	22	4	221	Chacornac	1856
DAPHNE	41	2.7691	.266	15	29	4	223	Goldschmidt	"
PALLAS	2	2.7696	.24	34	43	4	223	Olbers	1802
THISBE	88	2.7702	.165	5	15	4	224	Peters	1866
GALATEA	74	2.7777	.238	8	59	4	231	Tempel	1862
BELLONA	28	2.7785	.15	9	21	4	232	Luther	1854
LETO	69	2.7804	.188	7	57	4	223	"	1861
TERPSICHORE	81	2.8563	.212	7	15	4	302	Tempel	1864
POLYHYMNIA	33	2.8641	.339	1	56	4	809	Chacornac	1854
AGLAIA	48	2.8812	.182	5	0	4	825	Luther	1857
CALLIOPE	22	2.9107	.098	13	44	4	853	Hind	1852
PSYCHE	16	2.9237	.185	3	4	5		De Gasparis	"
HESPERIA	68	2.9717	.174	8	28	5	45	Schinparelli	1861
DANAE	59	2.9848	.162	18	15	5	57	Goldschmidt	1860
LEUCOTHEA	35	3.0066	.217	8	12	5	78	Luther	1855
PALES	50	3.0625	.287	3	9	5	150	Goldschmidt	1857
SEMELE	86	3.0908	.215	4	45	5	158	Tietjen	1866
EUROPA	52	3.0999	.101	7	25	5	168	Goldschmidt	1858
DORIS	49	3.1094	.077	6	29	5	176	"	1857
ANTIOPE	90	3.1188	.148	2	16	5	186	Luther	1866
ERATO	61	3.1297	.169	2	12	5	196	Förster	1860
THEMIS	24	3.1431	.117		49	5	209	De Gasparis	1853
HYGEIA	10	3.1511	.1	3	49	5	217	"	1849
EUPHROSYNE	31	3.1527	.22	26	27	5	218	Ferguson	1854
MNEMOSYNE	57	3.1565	.104	15	8	5	222	Luther	1859
FREIA	76	3.3877	.188	2	2	6	86	D'Arrest	1862
CYBELE	65	3.4205	.12	3	28	6	119	Tempel	1861
SYLVIA	87	3.4927				6	198	Pogson	1866
MINERVA	92							Watson	1867
UNDINA	93							Peters	"
AURORA	94							Watson	"
ARETHUSA	95							Luther	"
ÆGINA	96							Borelly	1868
CLOTHO	97							Tempel	"
	98							Peters	"

☞ MELETE was at first supposed by Goldschmidt to be Daphne, but was recognized to be a new planet on the calculation of its elements by Schubert in 1858. Hence its number is sometimes given as 56.

ELEMENTS OF THE MINOR PLANETS — Continued.

NAME.	Number.	Mean Distance.	Eccentricity.	Inclination of Orbit.	Sidereal Period.	Discoverer.	Date.
IANTHE............	99					Tempel	1868
HECATE	100					Watson.........	"
HELENA	101					"	"
	102					Peters	"
	103					Watson.........	"
	104					"	"
	105					"	"
	106					"	"

INDEX.

ROBINSON'S
Full Course of Mathematics.

No Series of Mathematics ever offered to the public have attained so wide a circulation or received the approval and indorsement of so many competent and reliable educators, in all parts of the United States, in the same time, as this.

Progressive Table-Book. This is a BEAUTIFULLY ILLUSTRATED little book, on the plan of *Object Teaching*.

Progressive Primary Arithmetic, *Illustrated.* Designed as an introduction to the "Intellectual Arithmetic."

Progressive Intellectual Arithmetic, ON THE INDUCTIVE PLAN, and one of the most *complete, comprehensive,* aud *disciplinary* works of the kind ever given to the public.

Rudiments of Written Arithmetic, for *graded* Schools, containing copious *Slate* and *Blackboard Exercises* for beginners, and is designed for *Graded Schools.*

Progressive Practical Arithmetic, containing the Theory of Numbers, in connection with concise Analytic and Synthetic Methods of Solution, and designed as a complete text-book on this science, for Common Schools and Academies.

The different kinds of United States Securities, Bonds, and Treasury Notes are described, and their comparative value in commercial transactions illustrated by practical examples.

A full and practical presentation of the Metric System of Weights and Measures has been added.

Progressive Higher Arithmetic: combining the Analytic and Synthetic Methods, and forming a complete Treatise on Arithmetical Science, in all its *Commercial* and *Business* Applications, for Schools, Academies and COMMERCIAL COLLEGES.

Particular attention has been given to the preparation of those subjects, which are absolutely essential to make good *accountants* and commercial business men.

The different kinds of United States Securities are described, the difference between gold and currency, and the corresponding difference in prices exhibited in trade, are taught and illustrated; also, a full Treatise of the Metric System of Weights and Measures has been added.

Arithmetical Examples. This book contains nearly 1,500 Practical Examples, promiscuously arranged, and without the answers given, involving nearly all the principles and ordinary processes of common arithmetic, designed thoroughly to test the pupil's judgment; to cultivate habits of patient investigation and self-reliance; to test the truth and accuracy of his own processes by proof; in a word, to make him *independent of a text-book, written rules and analysis.*

This work is not designed for beginners, but for those who have acquired at least a partial knowledge of the theory and applications of numbers from some other work, and it may be used in connection with any other book, or series of books on this subject, for Review or Drill Exercises.

An edition is printed *exclusively* for *teachers,* containing the answers at the close of the book.

New Elementary Algebra: a clear and practical Treatise adapted to the comprehension of beginners in the Science. The introductory chapter is designed to give the pupil a correct comprehension of the utility of symbols, and of the identity and chain of connection between Arithmetic and Algebra, leading him by easy and successive steps, from the study of written arithmetic to the study of mental and written algebra.

New University Algebra, containing many new and original Methods and Applications both of Theory and Practice, and is designed for High Schools and Colleges.

This book is not a *revision,* but a newly prepared and recently published work, thoroughly scientific and practical in its discussions and applications. It is a book filled with *gems,* and most of them original with the author.

Kiddle's NEW Manual of the Elements of Astronomy. Comprising the latest discoveries and theoretic views, with directions for the use of the Globes, and for studying the Constellations.

The Publishers offer this work to accompany "ROBINSON'S MATHEMATICAL SERIES."

The *plan* of the work is *objective;* the *illustrations* are new and *copious;* the *methods* greatly *simplified;* the *numerical calculations*, which are based on the recent determination of the Solar parallax, are made without recourse to any other than *Elementary Arithmetic*, and the *most rudimental principles of Geometry*

The book is designed for use in NORMAL SCHOOLS, ACADEMIES, HIGH SCHOOLS, SEMINARIES, and advanced classes in GRAMMAR SCHOOLS; and it is hoped that in this work the thorough and practical Teacher will find a *desideratum* long sought for in this department of science.

University Astronomy. Descriptive, Mathematical, Theoretical and Physical; designed for High Schools and Colleges. Large 8vo

New Geometry, bound separate, in cloth.

Plane and Spherical Trigonometry, in separate volume, cloth.

Concise Mathematical Operations. Being a Sequel to the author's Class-books, with much additional matter.

Key to Geometry and Trigonometry, Surveying and Navigation.

Key to Analytical Geometry, Differential and Integral Calculus, with some additional Astronomical Problems in the same volume.

Keys to the Arithmetics and Algebras, are published for the use of Teachers.

In it will be found condensed and brief modes of operation, not hitherto much known or generally practiced, and several expedients are systematized and taught, by which many otherwise tedious operations are avoided.

Brevity and perspicuity, two rare and commendatory excellences in a text-book, are leading features to this work, and, at the same time, the *rationale* of every operation, and the foundation of every principle, are fully and clearly shown.

The design throughout has been, not to *conceal*, but fully to *reveal* the difficulties of the science, and to encourage the learner, not to *avoid*, but to grapple with, and to overcome them; since, to the student of Mathematics labor rightly directed, is *discipline*, and discipline, after all, is the true end of education.

New Geometry and Trigonometry, embracing

Plane and Solid Geometry, and Plane and Spherical Trigonometry, with numerous practical Problems, the whole newly illustrated. New and original demonstrations of some of the more important principles have been given, and the practical problems and applications, both in the Geometry and the Trigonometry, have been greatly increased.

New Surveying and Navigation. With use of

Instruments, essential Elements of Trigonometry, Mensuration, and the necessary Tables, for Schools, Colleges, and Practical Surveyors.

The arrangement of the work, including as it does Trigonometry and Mensuration, requires that *two terms* should be employed in its completion, but students familiar with these subjects, by omitting them, can readily master the subject of *Surveying* proper in *one term.*

New Conic Sections and Analytical Geometry; prepared for High Schools and Colleges.

New Differential and Integral Calculus;

adapted for use in the *High Schools* and *Colleges* of the country—*thorough* and *comprehensive* in its character; and while it does not cover the whole ground of this branch of Mathematics, yet so far as the subject is treated, it is *progressive* and *complete.*

It is confidently believed that in literary and scientific merit, this work is not equalled by any similar production published in this country.